Communicating Mobility and Technology

Responding to the effects of human mobility and crises such as depleting oil supplies, Ehren Helmut Pflugfelder turns specifically to automobility, a term used to describe the kinds of mobility afforded by autonomous, automobile-based movement technologies and their ramifications. Thus far, few studies in technical communication have explored the development of mobility technologies, the immense power that highly structured, environmentally significant systems have in the world, or the human–machine interactions that take place in such activities. Applying kinaesthetic rhetoric, a rhetoric that is sensitive to and developed from the mobile, material context of these technologies, Pflugfelder looks at transportation projects such as electric taxi cabs from the turn of the century to modern day, open-source vehicle projects, and a large case study of an autonomous, electric pod car network that ultimately failed. Kinaesthetic rhetoric illuminates how mobility technologies have always been persuasive wherever and whenever linguistic symbol systems and material interactions enroll us, often unconsciously, into regimes of movement and ways of experiencing the world. As Pflugfelder shows, mobility technologies involve networks of sustained arguments that are as durable as the bonds between the actors in their networks.

Ehren Helmut Pflugfelder is Assistant Professor of Professional, Technical, and Scientific Writing in the School of Writing, Literature, and Film at Oregon State University, USA.

Routledge Studies in Technical Communication, Rhetoric, and Culture

Series Editor: Miles A. Kimball

This series promotes innovative, interdisciplinary research in the theory and practice of technical communication, broadly conceived as including business, scientific, and health communication. Technical communication has an extensive impact on our world and our lives, yet the venues for long-format research in the field are few. This series serves as an outlet for scholars engaged with the theoretical, practical, rhetorical, and cultural implications of this burgeoning field. The editor welcomes proposals for book-length studies and edited collections involving qualitative and quantitative research and theoretical inquiry into technical communication and associated fields and topics, including user-centered design; information design; intercultural communication; risk communication; new media; social media; visual communication and rhetoric; disability/accessibility issues; communication ethics; health communication; applied rhetoric; and the history and current practice of technical, business, and scientific communication.

1 **Computer Games and Technical Communication**
Critical Methods and Applications at the Intersection
Edited by Jennifer deWinter and Ryan M. Moeller

2 **Visible Numbers**
Essays on the History of Statistical Graphics
Edited by Miles A. Kimball and Charles Kostelnick

3 **Communicating Mobility and Technology**
A Material Rhetoric for Persuasive Transportation
Ehren Helmut Pflugfelder

Communicating Mobility and Technology

A Material Rhetoric for Persuasive Transportation

Ehren Helmut Pflugfelder

Routledge
Taylor & Francis Group
LONDON AND NEW YORK

First published 2017
by Routledge

2 Park Square, Milton Park, Abingdon, Oxfordshire OX14 4RN
711 Third Avenue, New York, NY 10017

Routledge is an imprint of the Taylor & Francis Group, an informa business

First issued in paperback 2018

British Library Cataloguing in Publication Data
A catalogue record for this book is available from the British Library

Library of Congress Cataloging-in-Publication Data
A catalog record has been requested for this book

ISBN: 978-1-4724-3472-2 (hbk)
ISBN: 978-1-138-31975-2 (pbk)

Typeset in Times New Roman
by Apex CoVantage, LLC

Contents

Front matter for existing citations

Portions of Chapter 2 and Chapter 4 are derived, in part, from a chapter entitled "Is No One at the Wheel? Nonhuman Agency and Agentive Movement" in *Thinking with Bruno Latour in Rhetoric and Composition*, published by Southern Illinois University Press in 2014.

Portions of Chapter 2 are derived, in part, from an article entitled "Rhetoric's New Materialism: From Micro-Rhetoric to Microbrew" published in *Rhetoric Society Quarterly* in 2015.

Figures

Acknowledgements

While this book was initially a very different dissertation project, and has been re-written several times, my dissertation committee has been invaluable in guiding the initial stages of the project. I wouldn't have attempted such an unusual technical communication project without the encouragement of Michael Salvo, who, when I first wondered what a rhetoric of transportation would look like, responded with, "keep asking that question." Pat Sullivan always lent a wise ear, Thomas Rickert pushed me to consider additional theoretical perspectives, and Jenny Bay helped me to think outside whatever box I stuck within. I have also had the opportunity to work under or take courses with a number of excellent faculty over the years who have also helped me find my career path: Irwin Weiser, Richard Johnson-Sheehan, Shirley Rose, Dave Blakesley, Linda Bergmann, Sam Blackmon, Tony Silva, Janice Lauer, Kurt Koenigsberger, Kimberly Emmons, Martha Woodmansee, Tom Bishop, Christopher Flint, Athena Vrettos, Bill Covey, Marnie Petray-Covey, Cindy LaCom, Robert Crafton, James Strickland, Derrick Pitard, and Sharon Sykora.

So many friends in rhetoric and composition have listened to the various stages of this research project, from those who simply heard the first emergent ideas, to those who listened as I worried my way through more Heidegger than was useful or appropriate. Some big thanks go to Cris Elder and Tyler Johnson, who hosted many dinners and were host to great conversation, Josh Prenosil, whose ideas are always cutting-edge, Megan Schoen, Kristen Moore, Liz Angeli, Jen Talbot, and Jeff Bacha. It was a great pleasure to go through a degree program with people as smart and enthusiastic as you all. I've always left conversations with Lars Soderlund, Kyle McIntosh, Caitlin Spronk, Don Unger, Allen Brizee, Mark Hannah, Sarah Harris, Jeremy Cushman, Laurie Pinkert, Tammy Conard-Salvo, Danielle Cordaro, Chalet Seidel, and Morgan Reitmeyer with a new outlook and new ideas.

My current colleagues at Oregon State University have also been a great help to this project, as they've watched it move from a recently completed dissertation to a finished book. Tim Jensen, Neil Davison, and Peter Betjemann have always been good listeners and great company on a bike ride. Lisa Ede, Sara Jameson, Vicki Tolar-Burton, Chris Anderson, and the other rhetoric and writing faculty have supported my work and made me feel at home in a brand new state and in a brand new department. I'm also lucky to work around such generous and thoughtful people such as Evan Gottlieb, Anita Helle, Steve Kunert, Christina Leon, Ray Malewitz, Rebecca Olson, Elena Passarello, David Robinson, Lily Sheehan, Chris Nichols, Emily Davis Malewitz, Joy Futrell, Jillian St Jacques, Felicia Phillips, Megan Ward, and Tara Williams. Thanks go out, too, to those graduate students who I have had the pleasure of talking with in the Thesis Writers' Meetings. Your insights have helped me understand my own work and teaching in countless new ways.

This work would not have been possible without the additional support provided by several different organizations. First, I should thank the committee for the Conference on College Composition and Communication for awarding my dissertation the Outstanding Dissertation Award in Technical Communication for 2013. Without their nod of approval, I don't know if I would have attempted to generate a book from this research. I received support in the form of a grant from Oregon State University's College of Liberal Arts and a very generous fellowship, including a wonderful new office, from Oregon State University's *Center for the Humanities*. I also know that this project wouldn't have been completed without the thoughtful support of Miles Kimball, series editor for Studies in Technical Communication, Rhetoric, and Culture, Ann Donahue at Ashgate, and the helpful and supportive feedback from the anonymous reviewers.

On a more personal note, my parents, Helmut and Elissa, have always supported my decisions, even though graduate school seemed to take me further and further away from Pittsburgh. Your strength always inspires me. My sister Heidi and my new brother-in-law Brian have always been there with a good cat story, and my new family, Tom, Lynne, Matt, Kim, Luke, and Chase, have been wonderful and loving. Most importantly, however, none of this work would have ever been possible without the continuous and wonderful support, advice, and love from my partner, Kristin Griffin. A wonderful writer and kind teacher, you bring joy into my heart every day that we're together.

Preface

When he was a visiting professor at the University of Texas at Austin, noted rhetoric and composition scholar James Berlin became engaged in a spirited debate with some graduate students. Berlin, along with Charles Kneupper, had been arguing for the close relationship between rhetoric and social constructionist critical theory to a group who seemed particularly recalcitrant to the idea. As Sharon Crowley explains, one frustrated student asked, "Well, does rhetoric make airplanes fly?", to which Berlin replied, "Yes, damn it!" (Crowley 585, qtd. in Lauer 1994). In the conversation that followed, Berlin and Kneupper used the "airplane question" as a way to explain the differences between hard constructivism and social constructivism and rhetoric's relationship to both.[1] As Fred Kemp tells the story, *he* was that obstinate graduate student, and Berlin followed up his ardent defense with a discussion about:

> how rhetoric created the invention of the airplane, the organization of airplane companies, the running of airlines, etc. [. . .] it was the first time I had really understood the notion of rhetoric and the social construction of knowledge as a creating and cohering force behind our material reality. [. . .] At the time it was a shock to my system. Now I understand what he was saying.
> (Kemp 2014, pers. comm.)

When I first heard this now famous anecdote, through Michael Salvo at Purdue University, Berlin's response struck me – not only in its simplicity, but also because it suggests a powerful conception of how rhetoric and materiality function together. I have reflected back on this anecdote often, each time thinking of the implications to suggesting something so extreme for technical communication scholarship. We can read Berlin's response to Kemp as an invitation to remember the power of technical documentation in recording, ordering, and making meaning in complex technical projects. We can value precise science and engineering work, though we can also consider how technical communicators are adept in facilitating the coherence of knowledge created by the subject matter experts we work alongside. We can also use Berlin's claim to remember the significance of persuasion, discourse, visual rhetoric, ethics, and politics in instances when technical projects have failed and planes have crashed, as illustrated by existing technical communication research on the space shuttle Challenger disaster. Yet, to suggest that rhetoric can make an airplane fly, to claim that rhetoric actually keeps airplanes in the air – and understand that claim as more than a metaphor – requires a different understanding of rhetoric. Throughout this book, I will be exploring one alternate version of rhetoric, a material rhetoric that is particularly suited for technology and transportation.

Instead of tackling how rhetoric might keep airplanes aloft, my focus here is on a different form of transportation, that of automobiles. I am interested in understanding how rhetoric – or more precisely, *persuasion* – makes automobiles move and keeps them in motion. In order to follow how this happens, this book employs a definition of rhetoric different than its traditional understanding as intentional, conscious acts of persuasion generated by people to convince other people. I am interested in persuasion as it inhabits both humans and nonhumans, occurs in largely unconscious, unintentional ways, and is at its most powerful when habitual. This form of persuasion is emergent in materials that we make use of everyday and could be said to comprise a significant portion of the rhetoric of everyday life. We can look for examples of this persuasive force in both the very early days of the automobile and today's attempts at post-car mobility. With such a conception of rhetoric, we are also able to locate and argue for advantageous positions for technical communicators in today's innovative transportation projects, both ones that are underway and ones yet to be generated. While I call the rhetoric that grounds this technical communication work a "kinesthetic rhetoric," I do not mean to supplant any existing conceptions of rhetoric, nor do I offer a kinesthetic rhetoric as a structured system of technical communication methods. Instead, I intend it to be a material theory that draws from Greek rhetorical concepts, actor–network theory, and recent rhetorical and technical communication research. Above all, a kinesthetic rhetoric is a rhetoric more sensitive to how design and movement with designed objects – *together* – can be persuasive.

Finally, I do not mean to suggest that Berlin would necessarily approve of the project that follows, nor agree with my answer to how rhetoric can make automobiles, instead of airplanes, move. However, I do believe that Berlin would see value in employing rhetoric to respond to the increasing awareness that internal combustion automobiles are a problem that requires a solution. From growing greenhouse gas levels to increasing traffic, the Western version of automobile mobility is not an ecologically viable transportation option for the twenty-first century. While many see the solution to this mobility coming only from technical and scientific fields, I posit that any answer must wrestle with the more nebulous forces of persuasion and that rhetorically grounded technical communicators can be key agents of change. This book sketches out concepts that we can use a kinesthetic rhetoric to think through and includes examples that illuminate some of the theoretical concepts presented in Chapter 2. The chapters that follow move from broad conceptions of innovative transportation design, to a close scrutiny of the interfaces where humans interact with transportation technologies, to how those interfaces are connected in logistics of movement, and finally to the ways humans and nonhumans alike perform the art of navigation. I follow these investigations with a larger case study of a failed innovative transportation project of an autonomous, all-electric vehicle network for a large college campus. While I am certain there are other possible answers as to how rhetoric can make vehicles move, or keep them moving, what follows is an attempt to back up Berlin's bold assertion with a material rhetoric for persuasive transportation.

Note

1 I am indebted to Joyce Locke Carter and Fred Kemp for their recounting of this event to me.

1 Persuasive transportation

In late October of 2012, Hurricane Sandy tore a destructive path through New Jersey and New York, before devolving into a major winter blizzard for Pennsylvania and many of the Great Lakes states. Sandy's immense size and record-low pressures, along with its coincident arrival with a cold front from the west, led some to dub it "Frankenstorm," a name indicating both the immensity and intensity of the weather event. The statistics concerning Sandy's devastation are sobering – in the US the storm can be blamed for 147 deaths, a power outage that affected eight million people, the release of 10 billion gallons of sewage, and over $50 billion in damages (Blake et al. 2013; Kenward et al. 2013). Sandy swamped, pulled apart, or washed away a significant number of roadways, completely submerged seven subway lines, forced the closure of scores of bridges and tunnels, and cancelled over 19 thousand flights (Mutzabaugh 2013). After the storm passed and cleanup efforts began, both New York Governor Andrew Cuomo and New Jersey Governor Chris Christie initiated systems of what some have called "odd-even" gas rationing, where the final number of a car's license plate determines which day the driver can purchase gas. Some baulked at the idea of such regulation, especially since New York City last experienced rationing in 1979, when the Iranian revolution caused widespread fuel shortages. This time around, instead of geopolitical instability instigating oil supply uncertainty, many suggested the cause was global warming and the emergence of more dramatic and severe weather patterns, a different kind of phenomenon also prompted by human activity. Though the gas rationing was largely over by Thanksgiving, it left an impression on the minds of many who experienced the frustration of not being able to rely on privately owned automobiles for transportation. Further, it served as a reminder that climate variation will continue to alter the transportation networks we too often see as impervious to change.

Humans living in the twenty-first century are going to encounter the following difficulties: depleting oil supplies, increasing air and water pollution, crumbling infrastructure, ongoing safety concerns, worsening climate change, and continuing political strife (itself resultant from any and all of these other concerns). It stands as no stretch of the imagination to suggest that transportation already plays a large role in these issues. Oil will run out, pollution will take decades to reverse, roads and bridges will cost billions to fix and maintain, and climate change will impact nearly all sectors of life, including future political uncertainty, economic instability, and military action. The challenges we face in the coming decades may be more substantial and deep-reaching than those we have previously encountered, and will likely ask us to be better problem solvers than ever before. Design theorists Horst Rittel and Melvin Webber explain these kinds of challenges as "wicked problems," that is, problems that cannot be "solved" by a new invention, but ones that deal with a wide array of technologies, material constraints, institutions, people, organizations, perspectives, ideologies, and habits. Though Rittel and Webber pass off engineering and scientific problems as more easily solved, they suggest that planning-oriented problems are wicked because of their complexity, instability, and variability (1973, 160). All of the big environmental problems we face, it appears, are on some level planning problems in that we must mobilize large numbers of people and things in order to mitigate their negative impact.

It will likely come as no great shock that the dominant Western (and increasingly worldwide) form of automotive transportation contributes to climate change and is therefore one of our current wicked problems in need of a solution. As it stands, the internal-combustion-engine (ICE) version of transportation, with its increasing pollution, crumbling roads, dwindling fuel supplies, rising political antagonisms, dangerous highway accidents, and isolating suburbanization cannot continue indefinitely, nor as some claim, survive much beyond the midpoint of this century. As Steffem Böhm et al. describe, as a way of moving about the world, ICE transportation is made possible through a continual "shoring up" of the structures that make such a regime of mobility possible:

> It depends on a range of contingencies for its continued success, including the ability of geopolitical intervention and dominance to secure access to oil, the ability of planners and traffic engineers continually to provide for the mitigation of chronic congestion, the ideological success in rendering thousands of human deaths annually as 'normal' and acceptable, the ability to overcome opposition to road building, the capacity to navigate the fiscal crisis of the state to generate sufficient funds to promote automobile use, and so on.
>
> (2006, 10–11)

However, these contingencies are not only "shored up" by industries and governments, but also by people who participate in these forms of mobility. That is, when we make the decision to drive our children to school, commute to work by car, or take a ride in the country, we participate in sustaining the argument for ICE transportation. Even when we do not make conscious decisions concerning how we want to navigate the world, we feel the deep-seated effects of the transportation decisions of other people who do.

In order to get a clearer grasp of just how negative ICE transportation has become, we can focus on just one issue: the cars that we drive to and from work and their relationship to greenhouse gases (GHSs) and climate change. According to the 2010 *US Department of State's Climate Action Report*, the US is home to at least 254.3 million passenger vehicles (18). Together with motorcycles, light-duty vehicles produce about 63% of the US transportation sector's GHG emissions, which, in turn, make up about 28% of *total* US GHG emissions (US Department of Transportation n.d.; US Environmental Protection Agency 2014). Globally, America is the source of 19% of CO_2 emissions from fossil fuel combustion, where American light-duty vehicles are responsible for close to 10% of worldwide CO_2 emissions (DeCicco and Fung 2006). Some reports hold global transportation technologies responsible for about 13% of all worldwide man-made atmospheric GHGs (Intergovernmental Panel on Climate Change 2007, 36). Yet, despite dire predictions of how the rise in GHGs affects our climate, we also stand in a historical moment where much of this damage can be mitigated. Mobilities scholar John Urry (2007) points to innovations in transport policy, new fuel systems, new construction materials, the deprivatization of cars, and deepening links between communication systems and transportation as possible innovations that could help arrest CO_2 emission and climate damage (280). Likewise, the 2014 *Mitigation of Climate Change Report* from the Intergovernmental Panel on Climate Change suggests a host of new developments in transportation technologies that could help greenhouse gases reach levels of stabilization able to slow global warming. These include, but are not limited to, more fuel-efficient vehicles, fuel cell and electric-drive technologies, modal shifts from road transport to rail and public transport, new land-use and transport planning policies, rebates for the purchase of high-efficiency vehicles, mandatory fuel economy regulations, and greater investment in attractive public transport facilities and non-motorized transport. Devising methods to enact such changes is seen by some as one of the major concerns of this century.

If we *cannot* devise meaningful transformations to our mobility problems, Urry argues that humanity appears to have one of three futures, each stemming from unsustainable transportation practices that will lead to the eventual degradation of our current economic, ecological, and political situations. The first possibility, a "dystopian barbarism of unregulated climate change," assumes that regional, ethnic, and tribal warlordism will come to dominate a world deprived of infrastructure, food, water, mobility, and the standards of living many Western cultures enjoy today (2007, 289). The second future, the "dystopic digital Orwell-ization of self and society," involves an increasing level of personal surveillance, data mining, biometric security, and the metaphorical, if not near-literal, fulfillment of a universal panoptic gaze (276). However, Dennis and Urry (2009a) suggest humanity has a third future, one that results in "self-reliant (and probably also semi-isolated) communities" sustained through principles of "localism, diversity and autonomy" (149). They note that "[a]lthough localism may be preferable to the other scenarios it seems less probable" (164). In order to achieve a future that escapes the former two scenarios, we need to realize new ways of navigating the world that are made possible through the development of significantly different forms of transportation. Simply put, the internal combustion engine vehicle and its attendant and surrounding networks of force and control are not sustainable in their current form.

We may believe that in order to confront the complexities of our deepening transportation problem, we must somehow *fight against* internal-combustion engine automobiles, their manufacturers, their infrastructure, and how they coerce us into specific forms of mobility. For some, the only ways to stop the negative effects of automobiles are to ban, tax, or shame people into not using them. However, confronting the issues concerning mobility is less a matter of righteous individual choice and more an issue of understanding *the forces that coerce us into movement*. In order to understand how transportation is persuasive, we need to first realize that while we are affected by regimes of mobility, we are always part of ongoing movement-based persuasive forces. This persuasion is subtle and we may not recognize our involvement, but it shapes the kinds of narratives we can enact and the ways we perceive the world. Further, these arguments are not only made by people, but by a host of others, including the technologies we design and use and the environments in which we use them.

Key to understanding what lies at the heart of mobility is recognizing how any transportation is the result of a range of persuasive forces that require the compatibility, coordination, and compliance of many interactions. The argument made by a politician, as she lobbies her peers to fund a new bridge construction project, is ultimately part of the same persuasive force made by gas prices, speed limits, and road networks, as they convey to us whether we should make the journey from home to work by car. The decision by traffic engineers to place speed bumps in a school zone is an argument about the increased caution drivers should take, and the communication between the road surface, a car's tires, steering column, steering wheel, and our hands persuades us how fast the vehicle should go. All of these interactions comprise a network of persuasion, one which sustains itself if the people and technologies invested in it continue habits that maintain the argument, but one which falters and becomes less convincing, less persuasive, if the major actors stop playing their roles. When gas prices skyrocket, people tend to share rides more, seek out public transportation, bike to work, telecommute, and find ways to alter their habits based on the lack of compelling argument for putting more gas in their cars. At this moment in history, we are witnessing an intense interrogation of ICE mobility and are more invested than ever in inventing and implementing possible changes. As such, my focus here is the kind of mobility most endemic in Western culture: automobility.

Automobility

"Automobility" is a term used to describe the specific kinds of mobility afforded by independent, automobile-based movement technologies, most often the personal automobile. Additionally, the term also describes the complex cultural, bodily, technological, and ecological ramifications of our dependence upon separate mobility technologies. Urry's most complete definition of automobility comes from "Inhabiting the Car" (2006):

> I use 'automobility' here to capture a double-sense. On the one hand, 'auto' refers reflexively to the humanist self, such as the meaning of 'auto' in

> autobiography or autoerotic. On the other hand, 'auto' refers to objects or machines that possess a capacity for movement, as expressed by automatic, automaton and especially automobile. This double resonance of 'auto' is suggestive of the way in which the car-driver is a 'hybrid' assemblage, not simply of autonomous humans but simultaneously of machines, roads, buildings, signs and entire cultures of mobility.
>
> (18)

The concept of "automobility" suggests that we are autonomous, able to move about freely, and dependent upon no one for the experience of driving. Such is the liberty car commercials depict – drivers alone on winding roads, under blue skies, able to decide for themselves where to go and whom to take with them. The reality, however, is suggested in the other sense of automobility. When we partake in automobility, we engage in a larger persuasive force that links us to many other systems, choices, and interactions. The understanding that people are enmeshed within many different systems within automobility allows us to make the case that automobility, as a structure of how we see the world, runs deep through our conceptions of what it means to be human, how we think about other people, how we conceptualize space, and how we value the things in our lives. This affective and emotional appeal can be quite intense. As mobilities scholar Mimi Sheller (2004) argues:

> In societies of automobility, the car is deeply entrenched in the ways in which we inhabit the physical world. It not only appeals to an apparently 'instinctual' aesthetic and kinaesthetic sense, but it transforms the way we sense the world and the capacities of human bodies to interact with that world through the visual, aural, olfactory, interoceptive and proprioceptive senses. We not only feel the car, but we feel through the car and with the car.
>
> (228)

In other words, because a system of mobility is comprised of people and things, humans and nonhumans, whenever we engage *in* automobility, we become part and parcel of the continuing argument *for* automobility. Any restructuring of automobility would need to affect the entire system, not just introduce one new element, like a gas-electric hybrid vehicle, into such a habitual, persuasive, systemic action. We can implicate automobiles, drivers, pedestrians, roads, parking lots, houses, suburbs, petroleum supplies, food availability, energy infrastructures, and many other unique objects, systems, habits, and symbols as part of what make up automobility (Urry 2004, 27). Automobility is thus our era's dominant form of mobility, coercing and enabling many different kinds of movement and reliant upon a vast array of technologies forces for its continued availability.

While the automobile is often conceptualized as a device that liberates people from the time constraints and restricted mobility options of public transportation, automobility carries with it its own limitations. The automobile can be a source of freedom, a " 'standing-reserve' to be immediately on hand, as [Martin] Heidegger characterizes the airliner on the runway," though the automobile also locks us into a managed system of assumed flexibility (Urry 2004, 119–120). Our schedules are often predetermined by automobility and our daily agendas are influenced by the supposed freedom the automobile affords. This coercion can be read as lingering forms of modernity within our postmodern environs. Even though many politicians and historians tout automobility as an intrinsic part of the great sweep of American progress that led away from World War 2, automobility can also been seen as *modernity par excellence*. Cultural critic Mark Dery (2006) explains:

> The car is a Second Wave totem: ever-present reminder of the assembly line that made industrial modernity possible, Ur-commodity at the heart of post-war consumer culture, essential ingredient in suburbia and the dereliction of the nation's inner cities, prime mover behind the strip malling of America [. . .] [At] a time when cell-phones, laptops, and the wiring of the world have made a mockery of time and geography, the car is a nagging reminder that we still haven't figured out how to zap our Darwinian luggage – the body – from here to there, in *Star Trek's* transporter.
>
> (228)

Like much from modernity, automobility has become a very insistent, very real material problem we struggle against. As Böhm et al. describe, the current regime of automobility is becoming increasingly "impossible *in its own terms*," as we see growing pressure placed upon every facet of the argument in order to make it compelling (2006, 11). It is likewise becoming progressively more fragile. As we see the instability of the networks necessary to support automobility (global politics, foreign policy, environmental disasters, climate change, supplies of nonrenewable resources), we realize the long-term unsustainability of the argument. Such is the position of automobiles within automobility: a technology within a regime of movement at once dangerous, embarrassing, and ubiquitous, but yet so deeply embedded within our habits we often fail to notice how powerful it really is.

Of course, though it may be tempting to place blame for social, environmental, and political ills upon the automobile itself, or even upon specific kinds of automobiles (as has been the feeling toward large sport-utility vehicles), the automobile is not the only problem, nor will changing the nature of the automobile solve our troubles. The "single-shooter" theory of technology does not hold for the detrimental impacts of ICE transportation. While it does appear true that "few products have had as powerful a role in shaping the way we live, for both good and ill, and so few strongly define and limit what options appear available for us," as an individual technology, the automobile does not alone shoulder the blame (Watchel qtd. in Soran 2009, 184). After all, as Dennis and Urry explain in "Post-Car Mobilities" (2009b), echoing Heidegger's thesis in "The Question Concerning Technology," "a car is not a car because of its physicality but because systems of provision and categories of things are 'materialized' in a stable form" (239). My own investigation addresses automobiles and automobility, but concerns how the fields of rhetoric and technical communication can help address these associated problems *and re-direct* our concerns over automobility into productive actions in a post-car future.

The post-car future

We can begin to see automobility in a more abstract way, and begin to question our current structure of mobility, if we think of automobiles as a kind of communication technology. Of course, there have always been communication technologies *inside* our automobiles. Currently, they feature stereos, satellite navigation, satellite radio, wireless internet, radar, cameras, sensors, video displays, DVD players, and touch screens that facilitate a wide range of Web-connected services. However, just because digital technology proliferates in our cars, that does not make them communication technologies. We can consider automobiles as a kind of media because they are information-laden forms whose structure, limitations, and abilities shape the kinds of meanings available to the people and things that interact with them. Marshall McLuhan compared an automobile's effects to that of home televisions and horses. He argued that "[t]he simple and obvious fact about the car is that, more than any horse, it is an extension of man that turns the rider into a superman. It is a hot, explosive medium of social communication" (1964, 241). We might indeed conceptualize automobiles as "socially realized structures of communication," according to Lisa Gitelman's definition of media (2006, 7). For Gitelman, these structures involve both humans and nonhumans and "include both technological forms and their associated protocols, and where communication is a cultural practice, a ritualized collocation of different people on the same mental map, sharing or engaged with popular ontologies of representation" (7). Though I do not discuss automobiles as explicitly "cultural" structures, but instead as agents in rhetorical structures, much of Gitelman's definition is useful. Automobiles are more than the material forms they take, but the social connections and protocols of movement they constrain and enable. Such an understanding also allows us to experiment with new technologies and interactions designed to help us see a beyond the car.

While a post-car mobility may seem possible only through a radical restructuring of technology and society, many projects have emerged in the hope of energizing new, compelling transportation habits. Right now, we are awash in these theoretical post-car technologies, ones that alter some of the most fundamental relationships fostered under automobility, including aspects of ownership, agency, navigation, collision-avoidance, interface interactivity, real-time information collection, social networking, design and manufacture, fuel sourcing, shared transportation, and interactivities with other methods of transport. Car-sharing programs like RelayRides and limited-term rental schemes like Zipcar and CarShare have blossomed, suggesting that for some, car ownership is less important than occasional car use.[1] Self-driving, fully autonomous automobiles such as those developed under DARPA's Grand Challenge and more recent creations by Google have shown us mobile futures that question who is at the wheel when we are in a car. Likewise, semi-autonomous systems have entered mainstream production vehicles in the form of self-parking controls, emergency braking sensors, and lane-change warning and correction systems. Satellite navigation devices find new routes across town, advanced digital dashboards help us to use less fuel, and many phones connect (nearly) seamlessly to our vehicles. We now have the option to drive automobiles with all-electric, electric-gasoline hybrid, bio-fuel, electric-assisted, or hydrogen fuel-cell power plants. Soon, pending mountains of legislation, we may be able to buy, build, and use an open-source, 3-D printed, all-electric autonomous vehicle that charges wirelessly, links data with our phones, and encourages vehicle sharing.[2] Certainly, some will dismiss such a statement, and rightly so – the likelihood that we will see such a contraption on the road in the next few years is doubtful – though aspects of this futuristic vision will make their way into our everyday lives (Salvo et al. 2010). By themselves, such innovations are not representative of a post-car future, but these new technologies could, if persuasive enough, usher in compatible changes in the host of other structures that make up a mobility regime. That many similar and compatible technologies are being actively pursued suggests both the instability of the ICE automobility and the potential for dramatic change in our existing transportation networks.

For two years I was a technical communicator involved in one such futuristic post-car design. This project, under development at a large, Midwestern land-grant university from 2007 until 2011, named the Electric Personal Transportation Vehicle (EPTV), involved the design and manufacture of an all-electric, autonomous transportation system centered on the creation of a "pod car" connected to an information and guidance network.[3] The initial concept for the EPTV came from a professor in the school of mechanical engineering and a course populated by senior-level mechanical engineering students. In 2007, instead of allowing students to invent their own projects, this professor proposed a thought experiment: what could transportation on the university look like in the coming decade, assuming that the school implemented the construction projects that were proposed in the 2009 version of a campus master plan. Crucially, in this plan traffic would be diverted by several blocks around the campus, with the goal of being more pedestrian-friendly. Students in the mechanical engineering course were asked to design an environmentally sustainable transportation solution that worked within a large central campus unable to accommodate the current volume of buses and personal vehicles. Their solution was a transportation network built around the EPTV (see Fig. 1.1). As a technical project, EPTV was quite an interesting post-car design exercise, primarily because it integrated several emerging concepts, including an autonomous (self-driving) propulsion system, guided by several forms of redundant navigation technologies. EPTV pod cars required neither rails nor tracks and were designed to be as flexible as possible with a minimum of coordinated infrastructure. They were powered by an electric motor with replaceable banks of batteries and designed to run for eight hours at a time before being recharged. In order to use an EPTV vehicle, riders would text

Figure 1.1 A CAD rendering of a design for the EPTV Club's pod car

Source: Author and EPTV Club

the EPTV system with their cell phone and the closest vehicle to the designated stop, moving in the same general direction, would deviate from its original route in order to collect the rider. Unfortunately, for a host of reasons that I explore in Chapter 7, we now have very little physical evidence that the EPTV project ever existed. There is no longer any pod car, senior design courses have moved on to other projects, and the EPTV Club has disbanded. It did not advance from a proof-of-concept to a prototype and ultimately failed to establish itself as a compelling form of transportation. In the past, I argued that the EPTV project failed because of insufficient goal setting, information management concerns, and organizational issues (Pflugfelder 2011). Those were some of the everyday problems that the EPTV Club worked through, yet those were smaller-level concerns. The EPTV project also did not become a realized and viable transportation network because it never materialized into a persuasive form of mobility – for both designers and potential riders. However, despite its failures, examining the EPTV case from the perspective of technical communication, rhetoric, and the interdisciplinary field of mobility studies will help us understand how innovative transportation projects have become and can be persuasive.

Mobility studies and technical communication

Most research into automobility and mobility does not come from within the field of technical communication but from social science scholars who work within a "mobilities paradigm." This research agenda initially took root in the UK and the Netherlands during the 2000s, has offered useful definitions of "automobility," as we have seen, and addresses the critical examination of society as based upon movement, disability, travel, and tourism. This mobilities paradigm does not assume that people are static by default, but instead incorporates their movements, mobilities, and communications-at-a-distance into an understanding of everyday life. For Urry, the starting point for mobility studies:

> is that the analysis of mobilities transforms social science. Mobilities make it different. They are not merely to be added to static or structural analysis. They require a wholesale revision of the ways in which social phenomena have been historically examined. As we have seen all social science needs to reflect, capture, simulate and interrogate movements across variable distances that are how social relations are performed, organized and mobilized. (2007, 44)

Mobility is not simply a part of an abstract theory, but networks of practice within which we are always immersed. Sheller and Urry (2006) argue that a mobilities paradigm has been influential for the social sciences because it confronts decades of "a-mobile" research and sedentarist assumptions, where "accounting for mobilities in the fullest sense challenges social science to change both the objects of its inquiries and the methodologies for research (208). As Tim Cresswell (2006) has argued, such sedentarist assumptions have existed in the social sciences for years. Cresswell finds that methods not attuned to mobility and motion have produced a rooted kind of "sedentarist metaphysics," a term he borrows from anthropologist Liisa Malkki. A sedentarist metaphysics assumes that people, as well as concepts, ideas, and objects, are in a "neutral" state when they are not moving, and sees mobility as a form of dysfunction (Cresswell, 26–32). By bringing mobility back into our theoretical methodologies and research methods, we acknowledge that physical movement is important to all dimensions of life and our critical analysis of those dimensions.

In contrast to the social sciences, most transportation scholars in engineering, logistics, information systems, accident analysis, and administrative policy tend to approach mobility problems with more positivistic assumptions, identifying which systems of transportation are most appropriate for planning initiatives and making networks of transportation more efficient. Urry has much to say about the sciences' contributions to mobilities research:

> To the extent to which transport and communication are researched, this mainly takes place in separate disciplinary fortresses with little interchange with the rest of social science. The study of transport concentrates upon the changing nature of transport systems and has deployed something of a technical determinism. It little examines the complex social processes that underlie and orchestrate the uses of such transports. There is we might say too much transport in the study of travel and not enough society and certainly not enough thinking through their complex intersecting processes.
>
> (2007, 20)

Some fields have failed to consider the complex interactions between vast transportation networks, human actions, and the ecological contexts in which these dramas unfold. Juliet Jain (2006) likewise argues that "transport academics and practitioners regard travel as a means to an end, i.e., accessing a destination activity, and propose that travel time is 'wasted time,'" and Stephen Graham (1997) describes how urban studies has often operated from a positivist perspective, treating space as an absolute object (79; 34). Happily, however, we appear to be better positioned to address mobility issues now than at any time before. There is, as Nigel Thrift contends, "a 'mobility' structure of feeling in the air" (1996, 259).

As evidence for this structure of feeling, we can point to a host of popular, persuasive texts about the state of transportation, many of which are critical to a degree unseen since early twentieth-century opposition to steam cars. In the mid-century, Vance Packard's classic *The Hidden Persuaders* (1957) exposed the world of consumer research and psychology in automobile advertising and John Keats' *The Insolent Chariots* (1958) satirized American automotive culture. Ralph Nader's deeply influential *Unsafe at Any Speed: The Designed-In Dangers of the American Automobile* (1965) accused the industry of putting "design" and "style" before tested, engineered safety and ushered in a focus on industry liability. Automobiles have played a role in theory building as well, as Roland Barthes used a Citroën DS to explain semiotics in *Mythologies* (1957/1972) and Marshall McLuhan used the automobile as an example of media in *Understanding Media: The Extensions of Man* (1964). In the last decade, we have seen a significant rise in the critical thought applied to our current mobility problems. Just to name a few, such arguments can be found in Donald Shoup's *The High Cost of Free Parking* (2005), Tom Vanderbilt's *Traffic: Why We Drive the Way We Do* (2008), Sperling and Gordon's *Two Billion Cars: Driving Toward Sustainability* (2009), Catherine Lutz and Anne Lutz Fernandez's *Carjacked* (2010), and Taras Grescoe's *Straphanger* (2012). Additionally, our attention has also been focused on the technology of the hybrid and electric car in James Billmaier's *JOLT!* (2010), Seth Fletcher's *Bottled Lightning* (2011), Margo Oge's *Driving the Future* (2015), and the twin documentary films from Chris Paine, *Who Killed the Electric Car?* (2006) and *Revenge of the Electric Car* (2011). Many of these works lay blame for our reliance on ICE vehicles and squandered post-car futures at the feet of the automobile industry, which has been only too happy to profit from technologies that support short-term profits instead of long-term sustainability. While popular works have been exploring current and ongoing transportation woes for some time, technical communication, as a field, has only recently been attentive to mobility concerns.

Though less attentive to transportation, technical communicators are quite adept at researching complex networks of people, technologies, and institutions. Technical communication scholars have come to understand how written and visual genres exist in networks of meaning, and how those genres coordinate with technologies to enable knowledge activities in complex working and learning environments (Russell 1995, 1997; Spinuzzi 2003, 2008; Slattery 2007). Further, they are engaged in analyzing how texts, images, media, objects, institutions, and ideologies influence each another and how agency is shared amongst these different actors (Swarts 2007; Potts 2008, 2009; Salvo et al. 2010). In many cases, they work in positions between subject matter experts (SMEs), technologies, and users, expressing these connections and influences through ethical, symbolic communication and working to articulate the particular needs of each. Technical communicators are often situated as advocates for users, working to keep the needs and desires of people at the forefront of how a technology or text is designed and articulated (Sullivan 1989/2009; Redish 1995; Johnson 1998, 2010; Sun 2006). Further, technical communication involves the employment of design strategies in document, information, interactivity, and interface creation (Schriver 1997; Mirel 2003; Salvo 2004; Potts 2013), in situations where risk communication to users is key (Sauer 2002; Nagelhout et al. 2009; Frost 2013), or in scenarios when users create their own material (Kimball 2006; Mackiewicz 2010; Selber 2010; Swarts 2015). Much of what technical writers do, then, whether in organizations or academia, involves analysis and problem-solving within information-rich environments that ask them to understand the complex relationships between people and things.

Johndan Johnson-Eilola has described these kind of problem-solving efforts as symbolic-analytic work, a term he borrowed from Robert Reich (1991) to describe activities accomplished "by a wide range of people in the burgeoning information economy. Symbolic-analytic workers, as the title suggests, manipulate symbols" (2005, 18). Though it may be easy to envision much of the work suggested above being done through textual production in networked spaces, I argue that it extends to where symbolic language is made manifest, transmuted into material bodies, enacted in space, and emergent in movement and habit. Sometimes this mutation happens literally, sometimes metaphorically – but always in the locations where the symbolic becomes material, material becomes symbolic, and many times in both directions at once. For example, the technical communication work I performed while in the EPTV Club included a range of textual, visualization, and oral communication tasks that were intimately bound to the material production practices of the group itself. If the club's work had progressed to a public testing phase, technical communicators could have been part of an iterative design process where user-input, as articulated through usability experience methods, directly impacted material design. By seeking out these kinds of interactions, technical communicators put themselves in position to give voice to a range of people and things and are, arguably, more able to recognize the changing natures of these complex relations. Considering the complex problem of our transportation future, technical communicators are ideally poised to articulate how transportation is persuasive, and should be involved in transportation projects where technologies are simultaneously being manufactured while emerging as culturally symbolic representations. Such projects require *symbolic-material workers* in addition to *symbolic-analytic workers*. Symbolic-material workers are those that seek out deeper understandings of how cultural assumptions are coded into material productions and how those designs are continually re-cast as culturally meaningful objects. Additionally, symbolic-material workers play a role in articulating connections between the cultural and technological while engaged in the process of design. They oversee and assist in the movement of knowledge between the symbolic and the material, working to create more sustainable interactions that ultimately benefit users. Technical communicators can fulfill these roles and can engage in critical, productive work helpful in making sense of the material persuasions that happen every day.

This shift toward an overt consideration of material persuasion and the symbolic representation in mobility technologies would be a new one for technical communication research. In the recent past, the field has addressed transportation either metaphorically or because the site was convenient to other research interests. That is, we have seen professional and technical communication scholars working in transportation sites and new media scholars interrogating virtual movement, but not necessarily confronting issues related to physical movement and mobility. Douglas Birch and John Fielder's *The Ford Pinto Case* (1994) concerns ethics, risk, and cost–benefit analysis and Clay Spinuzzi's *Tracing Genres through Organizations* (2003) addresses genre and activity theory in a professional workplace that tracks automobile accident data. However, we can also point to several recent studies that appear to value the role of mobility or transportation. Miles Kimball's "Cars, Culture, and Tactical, Technical Communication" (2006) understands personal, tactical forms of technical communication that happen outside of official, corporate representations. He analyzes communities surrounding two different car repair and design manuals in order to identify reappropriation and user–producer forms of documentation. Jason Swarts' "Mobility and Composition: The Architecture of Coherence in Non-places" (2007) approaches mobility from a different direction, looking at how mobile information technologies, like cell phones and PDAs, enable work to be likewise mobilized across time and space. Swarts studies how both physical and virtual movement creates what Marc Augé calls "non-places" (sites of temporary attention and transit) and affects veterinary students' writing strategies. Finally, several works by Jeff Rice situate the importance of movement in networked spaces. For example, his article "Woodward Paths: Motorizing Space" (2009) explores the value of networked taxonomies and what are called "folksonomies," local representations of spaces, structures, and organizations. Rice draws on Bruno Latour's actor–network theory and Paul Virilio's *The Art of the Motor* (1995) to argue that digital technologies can mobilize space in ways that reconfigure the relationships between people, the things they work with, and the spaces where coordinated activities occur.

Perhaps even more directly than these studies, the arguments I offered in "The Children of Aramis" (2010) with Michael Salvo and Joshua Prenosil function as a precursor for the chapters to come. We claimed that technical communicators are positioned to articulate the *chōra*, a concept from Greek theatre variously defined as audience, stakeholders, users, consumers, voters, and/or commuters. Specifically, we showed where the values of the *chōra* could be recognized within transportation technologies and explained how technical communicators could begin tracing "features of mass transit development [in order] to present the children of Aramis (a failed transportation project) as case studies for technical communication." We outlined a "place for the technical communicator in an emerging age of transparent, collaborative design – a posthuman age where technological artifacts and humans cohabit networks of communication" (247). If successful, technical writers could "gather voices from the technocultural chorus [. . .] collect these voices into representative media, and present features of the chorus for consideration in the design process (256). In this vision, technical communicators articulate the desires of the many actors in a network and do their best to represent all of the potential actors, human and nonhuman, in fulfilling a project's potential. In this sense, technical communicators can assist in assisting in identifying how a transportation project can be made persuasive for specific groups of users.

Persuasive transportation

The locations for the technical communication work described above can occur in the design of automobile dashboard interfaces, the scripting of logistic networks, or the user experience testing for physical and symbolic interactions. These locations for interaction are valuable in part because transportation activities throughout the world – as enacted by the coordination of people and technologies – are forms of persuasion. Transportation technologies do not control us, nor us them, but the impetus to use and continue using a mobility structure lies in the interactions that occur between people and things. For example, when I pick up my keys and walk to my garage, I can say that I am about to drive my car, but when I am on the road, negotiating signs, hazards, and other drivers, my own sense of agency becomes dissipated among the wide array of factors that influence my every vehicular gesture. The movement that makes up the "driver–car" assemblage cannot be pinned down to a single person, symbol, or automobile part, but instead results from a range of forces acting together. I am persuaded to drive to the store because I know there will be parking, that gasoline is not too expensive, and that the trip will be relatively quick. When on the road, in my car, these assumptions are reinforced and experienced as compelling forms of motion, because automobile travel often quickly and safely delivers us to our intended destination. These kind of movement-oriented persuasions are never merely the intervention of people into situations, nor the application of meaning to cold, dispassionate, objects, but a process that occurs whenever we move, with various technologies, things, and affordances.

Just as our movements are never without technological assistance, we are likewise never entirely removed from our sociotechnical situations. Tools and technologies constantly surround us, to the extent that we are never apart from their influence, as Bruno Latour explains. Latour argues that we are never without technology: "naked humans are as rare as naked cosmonauts. To define humans is to define the envelopes, the life support systems, the *Umwelt* that make it possible for them to breathe" (2008, 8). In these conditions, we are immanently part of and taking part in persuasive interactions and movements. The assemblage of the human driver, nonhuman automobile, and the attendant network of roads, traffic signals, social conventions, safety regulations, laws, and environmental conditions is one of the most compelling forces we can experience. Automobility has been one of the most consistently absorbing regimes of movement humans have ever encountered. A basic tenet of Latour's actor–network theory (ANT), that power and agency are distributed in complex ecologies of humans and nonhumans, also fits well with technical communicators' goals to articulate the persuasive impact of materials in these same networks. Latour finds that the terms "actant" and "collective" better representative of the complex interactions between humans and nonhumans. In *The Politics of Nature* (2004), Latour argues that an actant is anything that ". . . modif[ies] other actors through a series of . . ." actions (75). Actors can be any thing whatsoever, including humans, texts, automobiles, politics, organizations, weather, etc. Multiple actors functioning together function as an actant. Actants modify other entities in a given collection or assemblage, and when movements are performed, they not only originate from humans, but from "*an association of actants*" (1999, 182). In a sense, while I am partially responsible for my automobile's movements, I share agency for those movements with the huge number of symbolic and material conditions that make the car move, everything from oil refineries to laws regulating highway travel. Latour's work has been instrumental in identifying how these networks function, particularly his work in *Science In Action: How to Follow Scientists and Engineers Through Society* (1988), *We Have Never Been Modern* (1993a), and *Aramis, or the Love of Technology* (1996). John Urry, and others, have likewise explored the value of ANT for mobilities studies, building these assumptions into mobility methods and methodologies. As Urry states, paraphrasing Latour, "we have never been simply 'human', let alone purely social" and as such, humans and nonhumans mingle in strange imbroglios of power and affordance (2007, 45). This persuasion not only happens when people act upon things in the world, nor when technologies force us into accepting their built-in ideologies, but occurs through interaction, as the two meet. This persuasion can take different forms, too, depending upon the kind of vehicular interactions that occur.

When sitting in an autonomous car, like one that Google is currently developing, we can easily state that the car has more control over the act of driving than we do – the navigation technology is obvious and the lack of driver controls suggests the same.[4] We cannot touch a steering wheel, but instead allow the car to "steer itself." Google's autonomous vehicles employ a laser range finder, radars, cameras, and computer hardware that process over 1 GB of data per second, "see" in a 360-degree field, and correlate information to high-resolution Google Maps (Guzzo 2011, 18; Gross 2013). But when we drive our own vehicles, similar kinds of shared agency happen. I rely on my car to do an array of complex articulations I do not need to understand, nor even have knowledge of, in order to drive to the store. When driving, I feel as though I am at the center of a complex network of forces all working in temporary harmony in order to sustain my intended trajectory. While I am one of the key actants, I am hardly the only one. In *Staging Mobilities* (2013), Ole Jensen highlights how these many actants "stage" a typical morning commute:

> Imagine travelling (by car, bus, train or bike) to work and walking from the car park or station to your office. During your trip you must certainly have been involved in multiple interactions with fellow drivers and pedestrians. [. . .] Your situational mobility from the morning trip, moreover, has elements of your own choice, such as selected route, mode of transport, relaxed or aggressive driving, choice of seat etc. These elements are all expressions of a 'staging' with a relatively high degree of self-determination. But along the way your practices were modified by traffic lights, timetables, road design, traffic regulations, information systems etc., reminding you that there is a 'staging' going on from above as well.
>
> (2013, 23–24)

Seeing such networks of actants prompts our acknowledgment that agency is shared and distributed among the people and things that impact movement. Further – as soon as we engage with mobility technologies at all, we are playing a role in the way that transportation generates persuasion. We are as much a part of the persuasive force of transportation as the automobile, the price of gasoline, or the roads we drive.

In order to identify these persuasive assemblages and their relationships to movement and mobility, the following chapters answer the questions "how is transportation persuasive?" and "what might a rhetorical understanding of transportation mean for automobility, innovative technical projects, and the field of technical communication?" To clarify, when I question "how is transportation persuasive?", I am looking to understand how transportation itself is persuasive and how we can identify a rhetoric of transportation assemblages – I am not especially interested in the discourse about transportation and its rhetorical features, though my study is informed by such language. To clarify further, when I speak of "transportation," I mean "mobility technologies" – not only the individual buses, trains, and cars that take us to and from work every day, but the many and varied technologies we use in moving throughout our respective environments. While there exists numerous possible approaches to the questions I ask here, the path I take connects ancient Greek rhetorical concepts with current rhetorical theory, as a way to ground the work of technical communicators. This work relies on a material rhetoric in order to approach a speculative post-car future.

This chapter sets the stage for understanding how networks of people and things are persuasive and how research in both mobility studies and technical communication can enable productive analysis on and work within innovative transportation projects. Chapter 2 establishes some of the assumptions for a rhetoric more attuned to movement, what I am calling a kinesthetic rhetoric, as part of a larger theoretical positioning of the technical communicator as symbolic-material worker. I articulate specific attributes of a material rhetoric that we can use to consider issues of movement in general and material movement in particular. I establish that while many rhetoric scholars have defined persuasion as intentional, symbolic communication that originates from and can only be interpreted by people, recent theory suggests otherwise. Specifically, I focus on persuasion as being neither dependent upon humans for an origin, nor always intentional – it can emerge from nonhumans and manifest in habitual, unintentional, and unconscious ways. Bruno Latour's actor–network theory helps identify how humans and nonhumans are linked in their movements and aids us in questioning long-held beliefs about rhetorical agency. This conception of agency also challenges previous conceptions of the relationship between rhetoric and materiality and prompts a new understanding of how material actants can be persuasive. That is, nonhuman materials can be persuasive on their own terms, not only through human-oriented symbolic identification. From these fundamental theoretical claims, I posit *kinesis*, Aristotle's term for movement with limits, as a rhetorical term for persuasive movement, or what I call a kinesthetic rhetoric. This chapter grounds additional terminology I expand upon in Chapters 3 through 6, which investigate specific features of a kinesthetic rhetoric.

Chapter 3 uses the theory and positioning for technical communicators to begin a sequence of analyses of persuasive transportation. It begins with a short narrative about the East German Trabant automobile and how it was once appropriate for the restrictive political context in which it was first designed, though eventually ridiculed once the Berlin Wall fell and that context changed. This chapter concerns both design and the relationship between design and *techne* in particular. In short, I argue for strong connections between design science and "explicit techne," and design thinking and "abstract techne." Using these concepts, I show that both forms of *techne* are useful and necessary in the design of innovative technical transportation projects. Too much explicit *techne* can lead to formulaic designs, though too much abstract *techne* can result in unwanted conceptualization. Explicit *techne*-thinking can also lead to isolated work within a technical project and result in projects "designed by committee," like the famed Pontiac Aztek. Conversely, though innovative transportation projects may make use of similarly innovative online, open, or crowdsourced design tools, by themselves these tools do not guarantee success. This chapter argues that transportation projects need to consider *hyle*, a Greek term for the material in a given project, and work with that *hyle* throughout a situated design practice. For example, electric taxicabs in the early 1900s succeeded when they worked with the *hyle* of their design, but failed when they stopped. For technical communicators, a *hyle*-centered design can coordinate with existing user-centered design models and methods.

In Chapter 4, I move in closer to the relationships between human drivers and the nonhuman actants they encounter in automobiles. Specifically, it concerns the persuasive situation of the interface and how interfaces function, through metaphor, as rhetorical interactions. The chapter explores the design firm IDEO's work with the Ford Motor Company and their collaboration on a successful interface design as well as one that was less enthusiastically received. I argue that while technical communication and human computer interaction researchers have long been focused on the interface, recent work on technical communication has expanded beyond the graphical user interface (GUI). Employing a definition of interface initially developed for work in physics, I explore how transportation metaphors function as the persuasive result of how drivers make sense of designers' technical mediation work. While too many unique or novel interface metaphors can be confusing for users, catachretic metaphors are also necessary, especially with a concern for human drivers. Finally, showing how these definitions of interface and metaphor feature in Ford Model Ts, I explain how technical communicators can consider the value of metaphor in transportation projects, especially in the development of semi- or fully-autonomous technologies.

Chapter 5 begins with a story about what happened when the electric automobile company Tesla asked a journalist to review its network of charging stations. What resulted from the very public dust-up was a revealing look at how many Americans feel about logistics. This chapter works to dispel the notion that "logistics" are a-rhetorical or static sequences of material structures that support movement. While Aristotle identifies a human faculty for logistics, the Greek concept of *logistikos* allows us to see the relationship between logistics and the rhetorical canon of delivery. I identify logistics as a performed delivery, closer to more recent articulations of delivery as a "rhetorical velocity" or a strategy for circulation. When we engage in logistical thinking, for example, we distinguish trajectories through the world, identifying the kinds of interfaces and connections that must happen for the journey to take place. Some of this thinking is "logical," though much of it is not. The chapter closes with a critical analysis of the logistic developed for early electric vehicles, which was not particularly convincing, in part because it was dangerous, inconvenient, and depended upon too much variability in the interfaces users encountered. From this example, I argue that technical communicators can advocate for more environmentally sound, pleasing, and safe logistics options for users.

Chapter 6 focuses on the role of agency in the complex human–vehicular context of navigation, as part of a kinesthetic rhetoric. Beginning with a look at the history and development of cruise control systems, including the recent emergence of smart cruise control, I argue for a *techne* missing from Chapter 3. Specifically, I claim that a third form of *techne* exists alongside the explicit and abstract *techne* mentioned previously, though this *techne* concerns the *use* of materials rather than the *making* of materials. Building from Aristotle's work, Joseph Dunne identifies a user *techne*, perhaps closer to *phronesis*, though explicitly identified as an art. Aristotle's examples of this art include medicine, military combat, and navigation. This chapter concerns the latter and argues that a *techne* of ship navigation or chariot driving can be compared to modern-day vehicular navigation. I connect Greek myths about Athena's role in both design and use to the language of carpentry and ship navigation and identify where bodily knowledge (*metis*), propitious timing (*kairos*), and chance (*tyche*) all play a role in this *techne*. Further, I locate what was known as the "bit," or rhetorical and navigational agency, in Greek myth, so as to identify the role of agency in modern-day driving scenarios. This chapter uses these rhetorical features of navigation to locate a persuasive environment in vehicles, as they are being driven – relationships that we can likewise see in driver–vehicle environment (DVE) research. After exploring the DVEs of autonomous vehicles, I prompt technical communicators who work on similar projects to consider their work to be both user-centered and considerate of distributed agency.

The final chapter brings research offered in Chapters 3–6 to bear on one specific transportation project – the development of an all-electric autonomous pod car network at a large Midwestern university. This project, EPTV, was an active transportation design for a number of years, financially backed by the school of mechanical engineering, and supported by a robust student organization. Despite this support, however, the project folded in the spring of 2011, with little warning. I present this project as a case study and explore actor–network approaches to failed transportation design, including Bruno Latour's analysis of Aramis, before using several actor–network methods to understand the EPTV closure. After looking at the project from the perspective of a timeline, actor–network diagrams, and multiple definitions, I analyze EPTV from a kinesthetic rhetoric perspective, examining its design process, proposed interfaces, planned logistic, and assumptions about navigation and agency. Ultimately, I come to a different conclusion than Latour: he suggests that Aramis failed because of a lack of love, though I see "love" in terms of kinesthetic rhetoric and the persuasive power of a specific transportation network. Because of this, I also argue that technical communicators equipped with a kinesthetic rhetoric could have supported the project in various ways. Each of these chapters works to illustrate how transportation is persuasive and how a sensitivity to a kinesthetic rhetoric can assist in the design of innovative transportation projects.

There is no doubt that the future of transportation will require pioneering technical developments that allow us to see mobility in entirely new ways. That said, for the kind of sea change required to avoid some of the more dire predictions of climate change, any innovative mobility technology will also require a corresponding level of social acceptance. This book is not about how individual technologies can impact the production of GHGs, for example, but about how transportation technologies are persuasive and how any changes in our current regime of ICE automobility will require persuasion at the technical and cultural level. A huge array of humans and nonhumans need to function in concert in order for any network of mobility to function, which is precisely when and where we see persuasive transportation in action. Of course, if we want to make concerted, lasting change in our transportation future, we need to combine thoughtful, long-term analysis with sustainable design and production. Technical communicators, equipped as they are with rhetorical theory, practical knowledge, and productive skills, are uniquely positioned to understand and actively participate in these kinds of efforts.

Notes

1 Recent studies, such as those performed by KRC Research (and commissioned by Zipcar) claim that "millennials" feel that losing a cell phone or a computer would impact their daily lives more than losing a car (Zipcar 2013).

2 In 2014, Local Motors created the first 3-D printed vehicle on-site before the International Manufacturing Technology Show in Chicago. Printed in less than two days, Local Motors also hopes to sell this vehicle, called the Strati, to the public.

3 I have changed the name of the transportation project and omitted the name of the university, because many of the original participants in the project were undergraduate students. A portion of my research was approved by an institutional review board (IRB), though I was also part of the club and participated in the project.

4 The 2014 prototype of Google's self-driving car, often said to resemble a panda, has no steering wheel or brake pedal.

2 Rhetoric and movement

If we consider the quintessential way that automobility is present in most Americans' lives, it may be the trip to and from work we make five days a week. This journey is frequently made by car and typically lasts about 25 minutes, 10% of which is spent sitting in traffic (U.S.DoC 2011). The percentage of people who walk to work decreased in the second half of the twentieth century and the percentage of people who take public transportation and ride their bike has increased slightly in the first few decades of the twenty-first, though about 75% of people still drive to work alone. In order to understand why Americans make these choices, we might turn to an examination of where people live, where jobs are located, and what kinds of transportation options are available. Though most Americans have few public transportation options, many also choose to use automobiles, if only because convenience trumps cost. However, this chapter asks the larger question, "how can we understand automobiles as a material persuasion?" Specifically, I want to understand what makes cars persuasive by examining the material interactions that occur with automobiles when we use them. This chapter will set out some of the assumptions for a rhetoric more attuned to persuasive transportation, what I am calling a kinesthetic rhetoric, as part of a larger theoretical positioning of technical communicator as symbolic-material worker.

In order to understand the way in which persuasion functions in the technological and material context of the automobile, we need to consider it as a largely indirect force – one that we are compelled by over time. This kind of persuasion is often unnoticed by human drivers and depends upon a wide range of actors in order to function. Nonhumans play significant roles in this persuasion, though rhetorical theory has often ignored them, or assumed them to be only worthwhile for their symbolic value. It is much more difficult to understand these subtle forms of material persuasion on their own terms, that is, as more than *symbols* that are useful to *people*. In order to ground this material rhetoric, I explore the history of approaches to material rhetoric and work to understand how the field has identified nonhuman forms of unintentional persuasion. Focusing on *kinesis*, the Greek concept for material movement, this chapter shows how *kinesis* is related to *energeia* and suggests that *kinesis* operates as a central concern of any study of persuasive movement. While Chapters 3 through 6 identify more specific realms in which a kinesthetic rhetoric functions, and detail additional concepts for technical communicators working in innovative transportation projects, this chapter attends to the major assumptions of how human and nonhuman movement is persuasive.

Rhetoric and intention

In technical communication, both as a field of academic research and an industry practice, rhetoric is used as a heuristic theory that helps technical communicators make sense of specific communication interactions, strategize for continued exchanges, and produce documents and materials that respond best to those interactions. That is, rhetoric is frequently used to ground analysis and decision-making in various technical communication activities. Often positioned as a theory of symbolic interaction, rhetoric has been traditionally recognized as a guide to practical and ethical action and a guiding strategy for textual production. Technical communicators incorporate rhetoric to advocate for the roles of audiences and users, understand and modify genres, create and organize documents, analyze data sets, evaluate power dynamics in complex projects, recognize networks and patterns of articulation, comprehend and manage change, facilitate interpersonal communication, and grasp the rich contexts of professional settings, among many other activities. It would be no understatement to claim that the domain of rhetoric – in technical communication and beyond – is understood as how intentional, symbolic, and often linguistic expressions made by humans are encoded into texts or technologies and then interpreted and acted upon by humans. In this chapter, I build from the work of other rhetoric theorists to reject this definition as comprehensive and suggest how persuasion is neither the sole possession of people purposefully acting upon things in the world, nor an ideologically embedded influence waiting to be released onto unsuspecting users at the moment of use. Persuasion also emerges in the many unintentional, material interactions humans and nonhumans have with each other.

Much of the assumption that rhetoric is not limited to human-specific, symbolic intention relies on actor–network theory's claims that we can neither divide cognition from material–biological form, nor culture from technology. Instead of clearly defined subjects or objects, the world is made up of hybrid quasi-subjects and quasi-objects, which highlight the possible tendency toward subject-ness or object-ness of certain things, but refuse to be rendered into neat categories (Latour 1993a, 53). Bruno Latour developed his notion of hybrids, partially from Donna Haraway's conception of "monsters" – organisms that are not born, but "made in world-changing technoscientific practices by particular collective actors in particular times and places" (1991, 297). Monsters, or hybrids, are both and neither; both human and nonhuman, as they are comprised of both, yet neither at the same time. According to Latour, we encounter a wide array of these hybrids – "imbroglios of science, politics, economy, law, religion, technology, [and] fiction" – every day (1993a, 3). In order to illustrate these quasi-subject hybrids, Latour famously re-reads the bumper-sticker phrase "guns don't kill people, people kill people." In his eyes, a person with a gun is a fundamentally different kind of subject than one without. Either a gun-citizen or a citizen-gun, a person with a gun exists as another kind of subject and the gun as another kind of object: "[t]he gun is no longer the gun-in-the-armory or the gun-in-the-drawer or the gun-in-the pocket, but gun-in-your-hand, aimed at someone who is screaming" (179–180). We are always part of some kind of hybridity, too, because we are never fully isolated from the huge array of nonhumans that surround us. Accordingly, it is tempting to read Latour's articulation of a citizen-gun/gun-citizen in transportation technologies, as Tim Dant has done in his analysis of the "driver–car" (2004). Dant argues that the driver–car stands as one of the most powerful forms of this hybrid agent position, because drivers-with-cars generate a sense of being "geared" to the world, perceiving possible mobilities in order to navigate space (2004, 71). From these actor–network assumptions, we cannot simply consider technologies as part of a scientific or technological matrix that we encounter and then walk away from, unchanged. Fundamentally, what it means to be human depends upon the various interactions, incorporations, and assemblages in which we are enmeshed.

For my purposes, Latour's work helps articulate a new framework for rhetoric and technical communication – one in which we are able to see that rhetoric is not only a human domain nor only the study of symbol-based persuasion, also dependent upon human intentionality. A great deal of what can be defined as *persuasion* occurs beyond or beneath human intention. For example, how are we able to declare what has the most agentive control over a vehicle when it is in motion? What deserves the lion's share of credit for propelling my car down the motorway? My foot on the gas pedal, the pistons in the engine block, the smooth tarmac of the highway, or the oil refinery that produced the gasoline in the tank? Far too often, when we consider how persuasion happens, we only attend to intentional, human, symbolic origins. However, seeing as how we encounter such a wide array of material interactions in our daily lives, it seems a disservice to cut them out of rhetorical consideration.

The evidence that much of (what we consider to be) human consciousness takes place outside of our bodies has fueled several claims regarding the diffuse, material nature of agency and cognition. Theories of "technological unconscious," for example, assume that humans attempt to offload much of their cognition to our designed environments – "smart" surroundings that help make decisions for us, so as to free up other kinds of cognitive efforts.[1] Nigel Thrift has posited these locations of cognition-beyond-the-body as "spaces of anticipation" in designed environments that help set up patterns of unconscious, often-unintentional human habit (2008). However, even beyond a "technological unconscious," which projects human cognition outward, it might be fairer to consider these relations as a kind of "technological anteconscious": a diffusion of agency and cognition throughout the material world (Thrift 2008, 10). The networks of relation that carry intense anteconscious influence, like that of automobility, affect all people, even those that do not actively participate. For example, people that go "carless" must still contend with the widespread structures that support automobility. The rules of the road, and the structure of cities around those roads, affect cyclists, walkers, and those that take public transportation, too. I reference the anteconscious features of automobility not simply to show the constructed status of identity or subjectivity, or to marvel at the permeation of nonhuman agency, but to show that material environments that surround us are deeply persuasive and strongly affect our habits and movements.

Rhetoric, agency, and materiality

I want to articulate, briefly, a history of when rhetorical study has not accounted for the persuasive abilities of nonhumans and chart how the field has begun giving material objects their due attention. As such, what follows is a conception of rhetoric that identifies persuasion as it happens in nonhumans and humans, while avoiding being locked into human-centric assumptions. I attempt to focus on rhetorical concepts that embrace distributed agency and allow for material persuasion. As I have suggested, actor–network theory has been particularly good at questioning the assumed relationships between people and things and the assumption that we simply "make use of" nonhuman materials. As the growing list of citations attests, ANT now plays a larger role in technical communication research (Scott 2003; Hawk 2004; Spinuzzi 2005, 2007, 2008; Winsor 2006; Hart-Davidson et al. 2007; Slattery 2007; Swarts 2007, 2008, 2010, 2011; Fleckenstein et al. 2008; Potts 2008, 2009, 2010, 2011, 2013; Graham 2009; Rice 2009, 2012; Salvo et al. 2010; Fraiberg 2013; Pflugfelder 2015c), and we can identify a growing acceptance in rhetorical theory towards the nature of distributed agency. The nature and power of agency and the status and positioning of agents have long been disputed facets of rhetorical theory, from the ancient Greek rhetorician Gorgias's concern over the power of language in his *Encomium of Helen* (2003), where human agents can be at the mercy of an almost magical rhetoric, to Plato's allegory of the charioteer in the dialogue *Phaedrus*. For many, the modern argument emerges from Lloyd F. Bitzer's "Rhetorical Situation" (1968) and Richard E. Vatz's counterargument in "The Myth of the Rhetorical Situation" (1973). For Bitzer, material conditions and contexts invite a rhetorical response that comes into existence "for the sake of something beyond itself" (3–4). In this model, once an "appropriate" rhetorical response is identified, people can generate real change within a given situation. Vatz, on the other hand, focused on seeing situations *as rhetorical*, because they "obtain their character from the rhetoric which surrounds them or creates them" (159). This contrast opened up further discussion of agency as either emergent from the material or social situation or as part of representational symbol systems. It is certainly not radical to claim that much rhetoric and composition scholarship in the latter half of the twentieth century has been derived from Vatz's conception of agency. By the early 2000s, however, this perspective on the nature of rhetorical agency was changing.

The dialogue over agency at the 2003 Alliance of Rhetoric Societies conference, and the subsequent debate over the general consensus about agency in the pages of *Rhetoric Society Quarterly*, brought additional issues concerning subjectivity and the role of agent positions to light. Cheryl Geisler attempted to summarize the collective wisdom of that conference and express the dominant positions about agency discussed there. She felt that the most important conversations were about the rhetor's ability to act; most conference-goers, she claimed, gave the agent some level of efficacy for action (2004, 12). If agency is a theoretical illusion, as some suggested, then it was then a necessary illusion because the danger in recognizing agency as a complete illusion would possibly render the field of rhetoric irrelevant (16). Christian Lundberg and Joshua Gunn's counterargument (2005) came in the form of a critique of Geisler's assumptions and seemingly casual dismissal of Gunn's claims during the conference. They believed Geisler confused agent and agency while conflating posthumanism and postmodernism – and that such a conflation suggests agency is *mere* illusion (86). While Geisler claims situations may be socially and rhetorically constructed (see Vatz), her articulation of an agent is not constructed in the same way, or at least to the same degree – which allows for the "possession" of some *thing* called agency. Disagreeing with Geisler, Lundberg and Gunn find that agency should be considered akin to a possessing spirit – something that, depending on a range of conditions, can take root in different agents. Lundberg and Gunn's argument opens up the potential for agency to be freed of humanist confines and at least offers the prospective for nonhuman agents. Frustratingly, however, scholars' examples of agency are frequently human rhetors finding and articulating agency or otherwise emerging in/as the agent. There is very little discussion of nonhuman agents as rhetors, aside from indirect discussions on the sometimes powerful positions occupied by various technologies that affect our thoughts and actions.

In much the same way that rhetorical theory's conception of agency has changed over time, we are able to identify a shifting perspective in how nonhuman materials figure into persuasion. Throughout much of the history of rhetorical theory, scholars have been dismissive that material objects are able to contain and express their own persuasive force. When material forms are discounted and the realm of rhetoric is assumed to be only that of language and any attendant representational arts, we can then say that things are not "given their due." Richard Marback thoughtfully explains that we fail to give things their due "when we oppose persistence of object to insistence on interpretation" (2008, 52). That is:

> we fail to consider that any object of rhetoric – or for that matter anything rhetorical – might be more than a mirror reflecting some idea that precedes it, shines through it, and exists despite our best efforts to change it. We fail to recognize meaning and significance are in the moment, in activities of interacting with others with and through objects.
>
> (52)

Key to understanding material objects' rhetorical positioning is a different and more appropriate recognition of how things in the world (the things we make, use, and move with) are persuasive. Marback argues that these everyday interactions are deeply important, because:

> We all move through spaces of multiple scale everyday, often by means of manipulating objects. And we gesture as we occupy or move through spaces,

> with our gestures adjusting in response to perceptions of a space. Our gestures, our manipulation of objects, our occupation of space are activities in the present through which we engage the world with our bodies, bringing the features of both world and bodies into awareness of each other.
>
> (63)

Attention to the gestures that occur in physical space grounds rhetoric in the material by asking us to recognize the rhetorical expressivity of material forms. Of course, while I do not want to suggest that rhetoric does not happen in/through language, because such a claim would be foolhardy by any measure, I do want to add to our understanding of how rhetoric occurs in these material interactions, both with and without human engagement.

We would be remiss if we claimed that rhetoric has never considered the material in understanding the nature of persuasion. A great deal of rhetoric scholarship has considered materiality by way of the material aspects of rhetoric or the material field in which rhetoric occurs. As I have argued previously (2015b), this research might be best described as an attention to the materiality of rhetoric, though not the rhetoricality of the material. Barbara Biesecker and John Lucaites, in their collection, *Rhetoric, Materiality and Politics* (2009), historicize rhetoric's approach to the material as self-consciously emergent with Michael Calvin McGee's 1982/2009 article "A Materialist's Conception of Rhetoric" and his argument that rhetoric is a mode of consciousness that can be experienced. They see McGee's analysis of materiality as a prime example of "rhetorical materialism," where "the whole of rhetoric is 'material' by measure of human experiencing of it" (McGee, 23). Rhetorical materialism, in Biesecker and Lucaites' determination, is essentially a way to understand the social labor of rhetoricians as the mediating substrate of attempts to change and influence society (2009, 4). They suggest that:

> McGee insisted on our appreciating rhetoric as a mode of consciousness operating within an economy of phenomenological experience, quite literally the gestalt created by the complex interaction of speaker, speech, audience, occasion, and change at a particular moment in time. [. . .] Rhetoric, in other words, was not material in the sense of a "thing" like a rock or a tree, but rather as a palpable and undeniable social and political force.
>
> (2009, 3)

In this conceptualization, rhetoric is untethered from specific agents, but still safely within the realm of language, and the materiality which results depends upon people who use different language arts. Or, as Thomas Rickert explains, referring to McGee's essay, "materiality refers predominantly to rhetorical practice as material event" (2013, 21).

For their part, Biesecker and Lucaites are more interested in "rhetoric's materiality," a theoretical position that merges the rhetorical and the material. They suggest that attention to rhetoric's materiality emerged from Michel Foucault,

Jacques Lacan, Jacques Derrida, Jean-Francois Lyotard, and the poststructural focus upon a "new appreciation of the signifier's capacity to constitute a domain of reference rather than communicate meaning," where this domain of reference stands as more than signs, but the very materiality of the world (5). Here, language and symbolicity do not produce material contexts, because *symbol systems themselves are more than enough to offer up materiality*. Within this version of a poststructuralist frame, language constitutes all that can be conceived, and discourse produces speakers, audiences, contexts, and so on. Reality, and materiality, are thus both subsumed and defined through discursive articulation. Jeremy Packer and Stephen Crofts Wiley, in *Communication Matters: Materialist Approaches to Media, Mobility and Networks* (2011), offer an altered articulation of rhetoric's understanding of materiality from Biesecker and Lucaites'. They develop three different categories that encapsulate the field of rhetoric's approach to materiality. Their second category is likewise "rhetoric's materiality," though they describe this approach as concerned more with seeing materiality as rhetorical where material objects are seen to evoke responses, create affect, and promote or dissuade interaction in humans. Levi Bryant describes this perspective as the extension of symbolic representation, where the material object is "not an autonomous substance that exists in its own right, but rather a representation. As a consequence of this, all other entities in the world are treated only as vehicles for human contents, meanings, signs, or projections" (2011, 22). We invest symbol systems with material power, doing so (perhaps ironically) in order to recognize the value and importance of materiality. Further, a second complaint can be made against the "rhetoric's materiality" approach – that it not only assumes, but is locked into an ontological distinction between subject and object, despite claims to have surpassed this same division.

Packer and Wiley's description of a "third approach" is perhaps a little more tantalizing – an articulation of materialization that seeks to grasp the materiality of discourse itself that "crosses human-environment boundaries" and can include "rhetoric of (non)communication" (2011, 8, 206).[2] Here, rhetoric exists in "human-machine networks" where meaning can be articulated as "epiphenomena of a more fundamental, diffuse, and less human-centered process of composition" (Packer and Wiley 2011, 9). This third approach sees rhetoric as part of material processes, dependent upon, and part of, various material instantiations – as opposed to the study and performance of intentional, human-oriented, symbolic acts and only able to identify human-recognized, symbolic representations of reality. The assumption that rhetoric only applies to human-oriented symbolic acts is certainly understandable, because what else has linked the rhetorical tradition but a concern over communication, defined at times almost exclusively in terms of person-to-person written or verbal relations? Most rhetoric scholars do not shy away from humanism, but embrace the position of "big rhetoric" as a humanist meta-discourse, applicable to all realms of inquiry, able to absorb various other theories.[3] In this understanding, rhetoric is poised as a meta-discipline, perched "atop all the disciplines" so as to better recognize the interpersonal communication that shapes, colors, and generally influences other knowledge-producing discourses (Fleming 1998, 171). We might even argue that, historically, rhetoric has often understood power, materiality, emotion, affect, etc. solely from the perspective of symbolic persuasion.[4] It is certainly possible, however, to recognize how rhetoric could be better positioned to give material things their due respect and attention.

Rhetoric and nonhuman persuasion

In much the same way that Lundberg and Gunn's research troubles the human boundaries of agency, George Kennedy's previous work on animal rhetorics can be identified as one of the first modern identifications of nonhuman rhetorical agency. Kennedy's "A Hoot in the Dark: The Evolution of General Rhetoric" (1992) explores animal rhetorics, focusing at different moments on crows, chimpanzees, and leopards, to play out the implications of a speculation: what if rhetoric were abstracted to a kind of rhetorical force that anything could engage? Conjecturing that rhetoric is not a substance, exactly, he wonders whether some "might even seek to identify some quantitative unit of rhetorical energy – call it the "rheme" – analogous to an erg or volt, by which rhetorical energy could be measured" (2). While Kennedy suggests that task may be best left to the "experimentalists," he speculates what rhetoric would look like beyond language and speech (2). Kennedy argues this nonhuman, nonlinguistic rhetoric happens through the color of flowers, for example, as they attract insects and birds to assist pollination (10). Here, we are not in the alphanumeric world of human language production, but the world of nonhuman material persuasion. In general, this aspect of Kennedy's work was largely neglected when it arrived, possibly because few rhetoricians of the time wanted to work with the nonhuman interrelation he suggested.

Two current rhetoric scholars that carry on the experimental legacy of Kennedy's claims include Diane Davis and Thomas Rickert. Davis's work in *Inessential Solidarity: Rhetoric and Foreigner Relations* (2010) articulates a rhetoric of belonging and identification that exists before symbolic interaction and communication. She exposes a rhetorical response-ability, or relationality, especially with regard to new, interpersonal interrelations in which affective conditions provide the opportunity for or disclose identification. Before two people have verbal interaction, she argues, there is a wide range of affective and dispositional rhetorics that determine intentional rhetorical interaction. In showing that rhetoric "precedes and exceeds symbolic intervention," she finds that persuasion "frequently succeeds without presenting itself to cognitive scrutiny" (2; 36). While Davis' work seeks out the kind of rhetorical force that Kennedy believed could be engaged by animals, Rickert's work attends specifically to nonhuman, material persuasion beyond an animal or human context.

In *Ambient Rhetoric: The Attunements of Rhetorical Being* (2013) Rickert connects rhetoric with ambience, a move that allows him to see the "active role that the material and informational environment takes in human development, dwelling, and culture" (3). He builds off of work by Davis, Andy Clark, and Daniel Gross to identify rhetoric as "an emergent result of environmentally situated and interactive engagements, redolent of a world that affects us, that persuades us prior to symbolicity" (34). Rickert is adamant that rhetoric should *not be reduced to what is intentional or deliberate*, because large portions of human activity happen in the pre-symbolic, pre-conscious, or subconscious actions we engage in every day. Rickert connects his work with Latour's and provides a re-reading of Latour's speed bump (see Chapter 4). Latour argues that speed bumps exist somewhere between discourse and matter, where we "remain in meaning but no longer in discourse," and asks, "[w]here are we?" (1999, 186). Rickert's answer is that the speed bump "exemplifies material persuasion [and] equally connects to a network of relations that are rhetorical to the extent that an originary affect is already built in, awaiting catalysis" (2013, 208). This formation of rhetoric responds to a deficit in rhetorical studies, specifically how "rhetorical theory has yet to find a way to come to terms with symbolic mediation without occluding the material world except by recourse to standard theories of realism" (194). By no means am I suggesting that my claims provide such a solution, but this chapter identifies an alternate, albeit speculative, understanding of where a rhetoric intersects with materiality.

While much of my goal in this book is to understand how material persuasion occurs throughout vehicular technologies, I want to avoid being locked into the assumption that only humans can persuade. As a field, rhetoric seems better prepared than ever to speculate about new rhetorical formations that can attend to these forms of persuasion. Alex Reid's keynote speech at the 2012 *Computers and Writing* conference builds from many of the above assumptions, in order to make suggestions about new directions for rhetorical theory. He argues that identifying these new rhetorical formations is not about jumping on the latest theoretical bandwagon, but identifying how "to extricate ourselves from the anthropocentric symbolic action that has largely defined our discipline in the modern era" (n.p.). Reid explains:

> It is instead a realization that rhetoric was never and could never have been a solely human province. This is not a wholesale rejection of the theories and philosophy that have shaped our field but rather a recognition of their limits in addressing rhetorical challenges that we can no longer afford to imagine as simply discursive. From living and learning in digital communities to facing up to climate change, we must now compose rhetorics that incorporate technoscientific objects and build a future that includes them rather than divides them from us.
>
> (n.p.)

The rhetoric that Reid imagines would be speculative, insofar as it would explore relations beyond symbol-based confines. The promise of a material rhetoric is that it offers a way to examine the persuasive material forces and material agencies that exist in our world.

Few scholars have taken on the assumptions about rhetorical agency and nonhuman persuasion that keep us from recognizing shared agent positions better than Carolyn Miller – and her exploration of *kinesis* as a rhetorical concept is one of the most intriguing material- and movement-oriented rhetorics to be suggested in recent years. In her essay "What Can Automation Tell Us about Agency?" (2007), Miller surveyed teachers of composition and public speaking as to their opinions on a fictional product called AutoSpeech-Easy, software that when paired with a camera and microphone, offers autonomous, accurate assessments of students' speeches. Unsurprisingly, she found that automated assessment technologies (sometimes considered "robo-grading" systems) "cut to the heart of the theoretical and ideological concerns about agency" and ask whether posthumanist theory can "disperse agency to a machine" (145). She explains that most rhetoricians restrict the attribution of agency to humans because speech (the quintessential rhetorical performance) tempts us to focus on bodily performance, motion, and presence while recognizing agency as something that humans possess prior to a rhetorical act (147). The problem we have with performing for an automated assessment system "is not that it is inscrutable – audience is always inscrutable to at least some degree – but that we are unwilling to grant it such [agentive] presence and therefore cannot, in an important sense, perform" (149). In order to combat these notions, Miller argues that we should focus on agency as the *kinetic energy* of rhetorical performance, the energy of movement that occurs in performance.[5] If we see agency as a shared performance between rhetor and audience but ignore that nonhumans are likewise able to "have" agency, we will fail to see nonhumans as co-creators of agency. Miller is right that kinetic energy is created during rhetorical performance, however, I would add that we create the kinetic energy of rhetorical performance even when we are much less consciously aware of that performance.

We often assume that rhetorical performance involves forethought and planning, and for performances like premeditated speeches directed toward assessment software or live audiences of people, this is certainly accurate. However, there are countless instances in everyday life when we are unconscious of our rhetorical acts and unaware of how we are immersed in materially persuasive situations. If we consider rhetorical performance as also existing in mundane, everyday activities, our conceptions of agency change as well. When agency is only recognized as "potential agency," Miller contends, "it will be thought of as a possession or property of an agent (like a stationary stone), but if agency is a kinetic energy, it must be a property of the rhetorical event or performance itself" (147). Miller is neither harkening back to a Bitzer-like claim that agency lies wholly within a rhetorical situation, nor suggesting that agency is a human ability constructed in response to that situation, but arguing that agency, though movement, is collective. Everyday rhetorical performances involve shared agency, which can be observed through movement itself. Further, the observation and theorization of kinetic energy, or *kinesis* forms of motion, can be a key concept for recognizing distributed rhetorical agency in groupings of humans and nonhumans.

Kinesis and energeia

In this articulation of a material rhetoric more attuned to movement, my goal is not to reformulate rhetoric, nor provide an entirely new rhetoric that displaces thousands of years of study. Certainly, some of the assumptions explained above for a material rhetoric are incommensurate with existing perceptions of what rhetoric is and how it can be employed. However, we can identify conceptions of movement important to this material rhetoric – *kinesis* (κίνησις) and its partner *energeia* (ἐνέργεια) – both of which have roots in Aristotle's writing, though were not considered fundamental to his articulation of rhetoric.[6] My use of these concepts is not an attempt to offer a faithful representation of Aristotelian physics or rhetoric. I simply find that grounding a kinesthetic rhetoric in Aristotelian theory illuminates an alternate path for how movement could have been a more fundamental concern in rhetorical theory.

In *Physics*, *Metaphysics*, and *Nichomachean Ethics*, Aristotle distinguishes between two types of motion, *energeia* and *kinesis*.[7] One of the main forms of movement in Aristotle's work, *energeia* does not describe movement as we would typically understand it, but instead identifies a kind of internal emergence. Aristotle explains that *energeia* is actualized movement that is perfect, and goes on to describe it as movement without limit, such as seeing, understanding, and thinking (his examples, not mine). What distinguishes *energiea* from other kinds of movement is that it *includes the end of the movement within the state of being that is the movement*. Or, to put it another way, *energeia* movements can be said to "have been" at the same time they "are." For example, in the movement of "living well," a person can be said to have lived well while they are actively living well. As J. L. Ackrill explains in *Essays on Plato and Aristotle* (1997), an *energeia* is movement with "no upper or lower limit to the time it may occupy, and that it is somehow equally and fully present throughout any such period" (148). In living well, one can be said to have "lived well" at any moment up to the present. Further, there is no compartmentalized stage of living well that one must pass through in order to complete "living well." For reference, *energeia* motion can be considered more of a state-of-being than a series of physical movements in the world, though of course *energeia* can be part of more dynamic, material movements, too.

Distinct from *energeia* is *kinesis*, which Aristotle describes as imperfect movement with limits, such as learning, being cured, walking, thinking, and house building (again, Aristotle's examples). *Kinesis* is Aristotle's more general word for material movement, where "*kinesis* is itself an *energeia* in the wide sense, an actuality as opposed to a potentiality" (Ackrill, 142). Unlike *energeia*, *kinesis* motions cannot be said to have happened until they are complete. For example, one cannot be said to have built a house until the action of building a house is over. In other words, *kinesis* motions are often recognized through their manipulation of material objects or movement among things in the world. For Aristotle, the distinction between *energeia* and *kinesis* does not lie in the specific kinds of examples mentioned here, but in the essence of the movements themselves. To put it another way, the determining concern is not whether movement takes place among or with material forms, but whether or not movement has, as part of its own coherence, a limit. Essentially, *kinesis* motions have a limit, while *energeia* motions do not (Ackrill, 146).

However, Aristotle's categorical differentiations between *energeia* and *kinesis* are not as clear as they first appear, and his examples only muddy the water, especially as we consider unintentional or habitual movements. Critics have suggested that the only difference between Aristotle's two kinds of movement is that one includes a specific event or referent, like the "house" in the abovementioned example of "house building." However, Aristotle did not make the distinction between *energeia* and *kinesis* with such simple criteria as the inclusion of a finished product. Instead, he established two determining rules that categorize movement. Ackrill explains that rule one suggests "If Xing can occur not in a time, it is not a *kinesis.*" Rule two then adds, "if at every moment in a period of Xing it is true to say 'He has Xed', then Xing is [again] not a *kinesis*" (151). For example, "seeing" can be said to occur irrespective of a specific time, and for every moment during "seeing" we can also say that someone has "seen." Therefore, "seeing" is not a *kinesis*. That, I suppose, may seem like a complex way of differentiating between two kinds of motion, however Aristotle's definitions are not nearly as robust as he seems to suggest.

For example, we might find it strange that Aristotle classifies the movement of "walking" as a *kinesis* and not an *energeia*, because "walking" does not appear to fit either of the concepts we can associate with *kinesis*. Indeed, Aristotle offers us a belabored analysis of walking in *Nichomachean Ethics* and throughout book five of *Physics*, where he adamantly argues that it is *simply not an energeia*. Yet, the example of walking as *kinesis* is easy enough for us to problematize by Aristotle's own standards. If one's primary method of locomotion is walking, then walking functions as an ongoing, habitual, unintentional way of being with no fixed time or ending. Likewise, one can be said to "have walked" even while one "is walking." Walking can exist as a way of being in the world and as a general trajectory of movement. For example, I am (at least partially) characterized by my ability to walk; I am a walker; I am bipedal.[8] What is the distinguishing difference between walking as a way of being in the physical world and walking as a limited form of locomotion that has a clear beginning and ending? Movements, especially habitual and recurring forms of mobility, are difficult to separate into types that are endemic to being and those that are limited. While motions such as "living well" are not the same as "building a house," everyday forms of movement tied to everyday ways of being, like walking (and indeed driving), serve two categorical masters: both as *energeia* and *kinesis*.

We can also look upon Aristotle's categories of movement and see why he argued so strongly that walking is a *kinesis*, even though it appears to fulfill the criteria for both categories (as do numerous other movements). Aristotle distinguished between those movements/actions that can only function along with a being's innermost conditions from those that appear to happen in the world. Or, to put it another way, he distinguished between movements closer to what he believed was a kind of internal emergence from those concerned with demonstrable trajectories through space and time. Such differentiations in movement are part of Aristotle's larger delineation between the natural and the manufactured, between *physis* and *techne*. As Janet Atwill explains, Aristotle's "definition of *physis* is completely dependent upon *techne*. The mutually constitutive relationship between art and nature is part of the very structure of Aristotelian thought"

(1998, 85). Of course, and Atwill notes as much, such distinctions are also problematic, as the boundaries between nature and technology/culture shift throughout Aristotle's work and throughout Greek rhetorical and philosophical traditions. We should also note a relationship between Aristotle's categories of motion and his ontological assumptions – ones that divide the natural from the human-made. Further, in Martin Heidegger's reading, they also help show how movement is deeply meaningful.

Specifically, Heidegger found *kinesis* to be of great importance in interrogating Aristotle's ontology.[9] Both saw *kinesis* as the kind of complete movement in the world that follows from the potential for motion. Heidegger argued that as long as something is being worked on, as long as movement on or with things occurs and the thing itself is changing, *kinesis* happens. For Heidegger, *kinesis* is not a compartmentalized kind of movement – movement that occurs only once and when over, causes no change upon the world. As Heidegger explains, *kinesis* is not simply a kind of moving about, not "merely one state among others, but is an *essential determination.* Therefore *motion as a mode of Being is fundamental*" (2007, 142). Heidegger understands Being, the criteria or conditions by which any specific entity can show up, as comprised through motion. Fundamentally, motion always *is*. That is, for Heidegger, there is always something moving and there is no state that does not include motion, because things can only *be* if there also exists some variety of *kinesis* (147). *Kinesis* is not only made manifest through physical movement, but happens at rest, too, because rest is not the absence of movement, but an integral part of movement. Heidegger suggests, "[t]here is also *kinesis* even if all there is is something at rest" (147). *Kinesis* occurs when one is building a vehicle, for example, as the movements required in manufacture, but also while the vehicle is being driven, while goods are delivered, and while our lives are impacted by the assumption of certain of the kinds of *kinesis* being available in the world.

While *kinesis* can be identified within a deeper ontological structure, an ontology based upon movement is not necessary to study persuasive movement. However, understanding the importance of movement for a rhetorical approach gives us the ability to tackle unintentional, nonhuman, material persuasions. In part, this is because a rhetoric more sensitive toward movement would not only consider the role of movement in people, but seek out movement regardless of where it occurs. Because persuasion can take place in a range of different human and nonhuman actants, a rhetorical approach that looks for movement can avoid only thinking of movement in intentional, human acts. Of course, while movement is in no way synonymous with persuasion, a kinesthetic rhetoric approach relies upon the search for and the consideration of movement in any given context.

The term "kinesthetic rhetoric" fuses the Greek for physical, material movement, *kinesis* (κίνησις) and the Greek for sensation, *aisthēsis* (αἴσθησις). Some may see the term "kinesthetic" and recognize a reference to Howard Gardner's *Frames Of Mind: The Theory of Multiple Intelligences* (1983). Gardner locates kinesthetic intelligence as the ability to control one's own bodily motions and the capacity to handle objects skillfully (206). Movements that require mastery over one's own body, such as dancing or swimming, those that require the manipulation of objects, such as playing baseball or the violin, and those that are materially productive, such as artisan craftwork, are often included in this subset of knowledge.[10] While such a concept of intelligence is certainly important to understanding the relationship between movement and rhetoric, I have a different primary concern. A kinesthetic rhetoric, unlike a theory of multiple intelligences, is neither centered only upon the human body, nor understands rhetoric as human sense perception. In much the same way that Diane Davis argues that rhetoric manifests as a pre-symbolic disposition and Thomas Rickert identifies persuasion is an ambient force, here I locate persuasion as emergent in *kinesis*. A kinesthetic rhetoric is an approach to studying and intervening in the development of material movement. Such a rhetoric could be employed to investigate a range of human–nonhuman interactions, and the chapters that follow employ a kinesthetic rhetoric to identify persuasive transportation. However, I want to make it clear that *kinesis*, by itself, if not a rhetorical concept that we can apply to different transportation contexts in order to understand how each is persuasive. Looking for movement and cataloging the range of motions available to any given actor will not reveal how transportation is persuasive.

Kinesis as analysis

We can recognize the potential usefulness – and the limitations – of looking for *kinesis* movements in the material interrelations in vehicular environments. In such a context, an interesting question might be, "how persuasive is the interaction between a semi-autonomous automobile and its human passengers?" In order to answer this question, we could study a specially designed semi-autonomous vehicle like a Nissan Leaf that is equipped with a "Highly Advanced Driver Assist System." In 2013, Nissan obtained a license to test this Leaf on public roads in Japan, under larger zero-emissions, zero-fatality goals and Nissan's plan to produce commercially viable autonomous vehicles. The "Highly Advanced" version of the Driver Assist System assists drivers with: lane keeping, freeway exiting, lane changing, overtaking of stopped vehicles, decelerating behind traffic congestion, and stopping for red lights (Nissan 2013). These partially autonomous features allow the person in the driver's seat to avoid most traditional interactions with the vehicle. For most road conditions, this grouping of nonhumans could safely navigate through traffic in accordance with road laws. The person in the driver's seat is still able to take control of the vehicle in unusual or dangerous road conditions, or if the desire to change course arises.

The *kinesis* movements that result from this constellation of humans and nonhumans are wide-ranging and complex. We could see interactions as persuasive, insofar as they combine trajectories of movement within familiar systems of vehicular travel with low failure rates. Often engineers who design and test such systems define these parameters, though we could also observe the movements of the humans inside the vehicles. Do they read, drink coffee, or take a nap, putting implicit trust in the autonomous features of the vehicle?[11] Perhaps they wince, brace for impact, grab the wheel, step on the brakes, or, as one University of Michigan study suggested, simply vomit? (Sivak and Schoettle 2015). Shifting our attention to the nonhumans in the car, we might try to understand the specific relationships, for example, between traditionally used wiring and wiring harnesses and how they cope with the demands of more onboard computer equipment. We could observe and record the functionality of the vehicle with regard to how the electronic systems interact with GPS and function with short- and long-range radars, ultrasonic sensors, and camera systems. If these intense and detailed relations are successful, the vehicle will likely complete its journey with a minimum of unwanted drama. Or, if failure rates increase and these connections are unsustainable, we might understand something about the inter-relations between different developmental technologies.

That said, a method of inquiry based upon simple cataloging the movements in the autonomous car, humans included, would only explain so much. Engineers, human–vehicular environment (HVE) researchers, and usability specialists already have access to some of this data. Likewise, creating a list of all known motions that occur when an autonomous vehicle is in motion, or even a list of all "important" or "impactful" motions, will tell us only a limited amount about the persuasive context of those interactions. Basically, *no list of movements will explain how persuasion occurs* in such a complex assemblage. Even if we name it something esoteric like *kinesis*, movement alone will not provide a unique or especially helpful way of understanding complex and emergent mobilities (Pflugfelder 2015a, 193–194). *Kinesis*, while a useful concept in a material rhetoric, needs to be used alongside other rhetorical concepts to help researchers and practitioners better understand the design and use of such technologies. For example, an attention toward *kinesis* motions could assist in locating agency, cataloging important relationships or identifying problems between different actors in a driver–vehicle environment. When paired with other technical communication research and design methods, such as Byron Hawk's concept of post-*techne*, seeking out *kinesis* forms of motions can be particularly useful.

Post-*techne* technical communication

Because Western modes of transportation are often habitual and powerful in their effects, the field of technical communication could realize an opportunity to understand the relationships between the humans and the nonhumans in this movement. Though cataloging *kinesis* movements is not particularly useful, technical communicators in transportation projects should focus on the potential for symbolic-material work. Symbolic-material work is built from Johndan Johnson-Eilola's articulation of technical communication as symbol manipulation in complex textual economies, but also takes into account the explicitly persuasive material contexts of this work. However, symbolic-material work returns to some of the principles that Robert Reich and Johnson-Eilola argued against – that technical communication work should move away from supporting industrial production and move toward distributed, online, networked information management and articulation. Technical communicators should not jump in a physical assembly line or assume authority in product manufacturing, for example, but because of the intense relationship between technical communication work and fabrication, they can think of their work as focused on material persuasion, especially in innovative transportation projects. In part, this work strives to understand the complex relationships between symbolic language and material persuasion.

A good model for this symbolic-material work can be found in Byron Hawk's essay "Toward a Post-*Technê* – Or, Inventing Pedagogies for Professional Writing" (2004). Hawk advocates for a materially sensitive, posthuman role for technical communicators, arguing that *techne* is a flexible concept that can be attuned toward deeply contextualized forms of production. He defines *techne* as "the ancient combination of art and technology in productive knowledge" and post-*techne* as "the combination of technique, the technical, technology, and *techne* that is grounded in posthumanism" (389). Here, a posthuman *techne*/post-*techne* recognizes that techniques are not applied in the abstract of material context, but that "[e]very element contributes to the situation [. . .] and contributes to the learning, thinking, and understanding that emerges in that context and through that interface" (379). The productive work of post-*techne* "should set up constellations of relations that allow its users to see something as something else-that is, to see in a new way through those constellations of relations" (379). Hawk continues this line of reasoning into the technical communication classroom and locates a positioning for technical communication teachers, ultimately for the benefit of students who will be expected to operate in complex workplaces and will need to "get a sense of how to develop techniques that operate in those specific kinds of situations [so as to] anticipate the conditions of possibility of that enaction" (384). A sensitive post-*techne* technique, then, would be more attuned to the ambient and context-dependent concerns in ecologies of production. In much the same way, technical communicators involved in innovative transportation projects operate in constellations of institutions, posthumans, technologies, and the very material of design itself.

Further, a post-*techne* is an art that assumes distributed agency. It recognizes that the "source of change or motion is always different and always split between producer and user; in other words, neither producer nor user is capable of determining productive knowledge" (Atwill 1998, 175). A post-*techne* is not a process that we apply to a given situation or context, but one that must be employed between humans and nonhumans in "ecological contexts in ways that reveal modes for enacting that open up the potential for invention (Hawk, 384)." What we have in a post-*techne* is a mode of production attuned to the ambient forces within a constellation of forces, bodies, and things. This is essentially the position of a symbolic-material technical communicator – someone who has the relations of humans and nonhumans as their main concern and articulates them as appropriately as possible. This focus is especially useful for innovative transportation projects, where material persuasion is difficult to identify, yet plays such an important role in success. In identifying this symbolic-material work, I do not mean to overturn Johnson-Eilola's suggestion for the value of distributed, networked, analytic work, but only add that technical projects often require a different positioning of technical communicators as advocates for material persuasion in its myriad forms.

This chapter has attempted to lay some of the theoretical groundwork for understanding how transportation can be persuasive by exploring the possibilities for nonhuman agency for persuasion and connecting ANT with rhetorical theory. I offered a short history of rhetorical approaches to material persuasion, identifying both obstacles to understanding how nonhumans can be powerful actants in generating persuasion and research that takes seriously the persuasive capacities of material forces. In part, moving beyond theoretical models that assume a human frame of reference helps us get closer to recognizing the capacity of materials to persuade. I articulated one approach for the study of unintentional, nonhuman persuasion, and suggested that the study of *kinesis* forms of movement could avoid the humanist assumptions that previous scholars upheld. While *kinesis* is certainly not synonymous with persuasion, it provides us with a material concept for a rhetoric that can aid us in interrogating mobility technologies and how they generate, and are generated by, persuasive assemblages of humans and nonhumans. The following chapters, 3 through 6, attempt to elaborate on a kinesthetic rhetoric, a material rhetoric sensitive to movement and applicable to the study of persuasive transportation. This methodology has the potential to become significant, as innovative transportation projects seek to take up new roles in everyday life and as designers create mobility technologies they hope are successful. These chapters also position technical communicators as workers equipped to understand the design of persuasive transportation, work that articulates the relationships between the material and the symbolic. The next chapter focuses on these design processes and the relationships between rhetoric and design, which re-connecting with the concept of *techne* discussed in this chapter.

Notes

1 Much of this work is beholden to Patricia Clough's *Autoaffection: Unconscious Thought in the Age of Teletechnology* (2000), where she understands the ontological relationship between the body and the machinic through a feminist psychoanalytical reading.

2 Packer and Wiley cite the origin for their third rhetorical approach to materiality as far back as Foucault's attempts in *Archeology of Knowledge* (1972) and twentieth-century continental media theory. They suggest that the goal of such work was, in part, to illuminate how specific material conditions made possible specific discourses. Packer and Wiley argue that this third approach avoids locating material forms in symbol systems or discourse, and seeing materiality as a document, but instead acknowledges that rhetoric exists in the material world as much as the material world expresses itself in rhetorical terms.

3 In Alex Reid's keynote talk at the 2012 *Computers and Writing* conference, he stressed that the field obviously operates within this framework, "focusing conventionally on symbolic behavior with the premise that we do not have direct access to the world referenced by the symbols" (n.p.).

4 Quentin Meillassoux explains similar representations as "correlations" in *After Finitude: An Essay On The Necessity Of Contingency* (2008). He argues, "[b]y 'correlation' we

mean the idea according to which we only ever have access to the correlation between [human] thinking and being, and never to either term considered apart from the other. We will henceforth call *correlationism* any current of thought which maintains the unsurpassable character of the correlation so defined" (5).

5 The prejudice against movement, and the assumption that our sites of analysis are somehow sedentary, has not been nearly as predominant in rhetorical theory as in other academic fields. Nigel Thrift, in his book-length study of motion and theory, identifies this prejudice toward movement as having slipped in with theoretical assertions about the lived experience of the moving body. These claims assume "that bodies are bodies-in-action, able to exhibit a kind of continuous intentionality, able to be constantly enrolled into activity," when, in many cases, we are neither actively attendant nor fully in control of our lived bodies (2007, 10).

6 All references to Aristotle are taken from *The Complete Works of Aristotle*, editor Jonathan Barnes.

7 In *Physics*, Aristotle explains that there are *three* kinds of movement, "accidental" motion, motion of part of a whole, and motion of something "in virtue of being itself directly in motion" (224a26–27). I see this delineation as only a further breakdown of *energeia* and *kinesis*, Aristotle's dominant categories of movement.

8 In considering walking, I in no way mean to suggest that people who are differently abled are ignored or unnatural with respect to movement. I have chosen to discuss walking because it is Aristotle's focus upon that form of movement and because it allows for an example that is often habitual and unconscious. Other forms of motion would fit this critique as well.

9 Martin Heidegger was convinced enough of Aristotle's valuation of movement to read him as a very early phenomenologist (Feenberg 2004, 4). Reflecting upon Heidegger's lectures on Aristotle, Andrew Feenberg claims Heidegger's reading recognizes that "Aristotle's greatness lies in having placed movement, *kinesis*, at the center of philosophical reflection" (28).

10 N. Katherine Hayles also employs the term "kinesthetic" involvement in *Writing Machines* (2002), where she offers a definition similar to Gardner's definition of bodily interaction. Describing the ways in which electronic texts signify in more engaged ways than print, she identifies "sound, animation, motion, video, kinesthetic involvement, and software functionality, among others" (20). Hayles' work addressed the kinds of reading that happens with electronic literature, which manifests in a "more complex activity actualizing the textual space so that it is kinesthetically vibrant as well as verbally resonant" – a view shared by Dene Grigar (Hayles 2002, 99; Grigar 2005, 105).

11 A 2014 study from the University of Michigan Transportation Research Institute suggests that while 56.3% of US respondents are positive about fully autonomous cars, 26.1% are "very concerned" about riding in such a vehicle (Schoettle and Sivak, 7; 11). About 35% of US respondents claimed they would "Watch the road even though I would not be driving" and an additional 23% would refuse to ride in a self-driving vehicle (18).

3 Design

Transportation designs are perhaps one of the easier locations where we can identify the assumptions that designers have made about who will use the technology, in what context it will be employed, and how it will convey people throughout the world, among other concepts. In the following story, we can see these assumptions at play and recognize how the design of transportation can be an insight into the material context of persuasion. In circumstances like that of the East German Trabant automobile, when the context was altered by other actants, some deep insufficiencies in the design emerged. The materials used in the design were not changed, but they became less persuasive. In one material context, the Trabant's body panels were a durable, upcycled design, though in another, they became both a nuisance and an environmental hazard.

Duroplast design

The Trabant's story begins in the decade after World War 2, during the most difficult rebuilding period for the newly formed state of East Germany. At the time, one of the many vexing questions for East Germany involved what their manufacturing facilities should build, that is, those facilities that were not destroyed in the war. Wolfsburg, in West Germany, had successfully rebuilt a Volkswagen plant and began manufacturing the Beetle, modifying an existent pre-war design. East Germany had the desire to manufacture a competitor to the Beetle, as both the Horch vehicle company and older Audi manufacturing facilities were in Zwickau, East Germany. The newly named *HQM Sachsenring GmbH*, under the larger manufacturing concern *Industrieverband Fahrzeugbau*, began to experiment with designs for a small, do-it-all family car. Their goals were similar to Volkswagen's – to produce, in large numbers, an affordable, durable, easy-to-repair vehicle. However, their efforts confronted the manufacturing difficulties inherent to their political and economic postwar situation – namely, steel was limited in supply and expensive to import, yet seemingly necessary to create automobile bodies. The vehicles Sachsenring produced during the 1950s, tractors and pre-war designs from the manufacturer DKW, could only be made in small numbers and were priced far out-of-reach of most of the population. The solution required a sea change in the material design of the vehicles themselves.

While East Germany may have been in short supply of automotive-grade steel, it did have other material reserves, including byproducts from large-scale manufacturing occurring throughout the countries now occupied by the USSR. The existence of glass fiber-reinforced polymers, or fiberglass, had been known since the late nineteenth century and had been developed during the war as a lightweight alternative to steel. Sachsenring engineers and designers built upon the principles of fiberglass and created Duroplast, a plastic resin strengthened by cotton. The material was comprised of a phenol, a result of the dye industry, and cotton wastes, which were shipped in from other states in the USSR (*Scientific American Frontiers* 1994). Under high heat, the materials were layered together for strength and then molded into shape – resulting in something not unlike Bakelite (which is make from phenolic resin and wood flour). Though it may not sound particularly robust, Duroplast was an effective car body material. It was lightweight, resisted dents, and completely rustproof. The use of Duroplast also made Sachsenring the first manufacturer to produce and sell a passenger automobile with a body made out of either recycled or upcycled materials.[1]

The Duroplast vehicle that Sachsenring first produced in any volume was the AWZ P70, or Zwickau P70, a monococque, transverse engined, front-wheel drive, sub-compact. Production eventually shifted to the P50, and eventually a P60 model. The "P" stood for "plastic." More people will recognize this vehicle, and especially the longest running model, the P601, as the Trabant, or "Trabi," a vehicle that carried many East Germans into West Germany after the Berlin Wall fell (see Fig. 3.1).

Figure 3.1 A typical P601 Trabant

Source: Sneldon 2010

The Trabant had an iconic shape, with two round headlights, and the two-stroke engine had few moving parts, simplifying repairs. Over three million P601 Trabants were produced between 1964 and 1991, easily making it the most popular vehicle in East Germany. In many ways, the Trabant could be called a success: the advanced composite body was a response to the material conditions in which the vehicle was designed and produced, they were easy-to-fix, and Trabants were reputedly safer than other subcompacts (Childs 2013). Trabant repair manuals were also unlike those created for Western European or American vehicles of the era. They clearly depicted "every single gasket, valve, hinge and so on of the car itself, in such detail that with a bit of determination and skill any ordinary East German would be able to take their Trabant apart and reassemble it" (Rubin 2009, 36). Further, updates to new Trabants could be retroactively applied to older models, so that even 15-year-old vehicles benefitted from design advances (such as they were). In fact, the average Trabant was on the road for an average of 28 years – much, much longer than a steel-bodied car of any era (*Scientific American Frontiers* 1994). When the Trabant debuted in 1964, it was a *relatively* advanced vehicle, yet it is now known as an object of ridicule, in part because of this longevity.

One of the major challenges in designing a car to be simple, easily updated, and durable is that large-scale innovations are nearly impossible, as they would necessitate a radical change in design. For the Trabant, major design changes were difficult – both from a design standpoint and from the perspective of officials within Trabant's parent company, who wanted to keep production lean and cheap. For example, Trabants initially used a two-stroke engine (a valveless engine with both intake and combustion happening in one revolution of a piston), which is durable, has fewer moving parts, and features a higher power-to-weight ratio than traditional four-stroke engines. Two-stroke engines require a mixture of oil and gasoline, and because Trabants were produced in large numbers, most East German gasoline stations offered pre-mixed Trabant fuel at the pump. However, this same two-stroke engine was also extremely polluting. The Trabant reportedly produced as much as 10 times the pollution of a modern vehicle – a situation that caused problems when East Germans drove their vehicles into West Germany.[2] Trabants also lacked a fuel pump, which meant that fuel was fed to the carburetor by gravity. Accordingly, the gas tank had to be located higher than the engine, which increased the risk of fire in an accident. Some of these problems can be blamed on the design's durability and Sachsenring's insistence on one sustained design, yet the most significant problem with the Trabant's design came about directly because of the fall of the Berlin Wall.[3]

When East German families loaded their Trabants and met up with relatives living in the rest of Europe, they realized that their cars were very much behind the curve, even when parked next to vehicles from other Eastern bloc countries. Comparatively, Trabants were loud, smelly, and small. Overnight, the value of a Trabant dropped from as high as 16,500 East German Deutschemarks to the rough equivalent of 50 cents, as Trabants were abandoned *en masse* by the side of the road (Mantle 1995, 144; James 2007). Therein lay another significant issue – because of this new context where longevity was much less desirable, the Duroplast body became a liability, not an asset. Unlike steel, Duroplast cannot be easily recycled. Made out of already upcycled materials, Trabant bodies could not be crushed in a typical car crusher (Duroplast fails to compress), burned in an incinerator (Duroplast gives off toxic fumes), or buried in the ground (Duroplast leaches chemicals). What had never been a problem for *East German* engineers was now a problem for *German* engineers. Apparently one solution is to shred the Duroplast, mix it with concrete, render the chemicals inert, and then press the material into bricks for building, though this was never accomplished on a significant scale (*Scientific American Frontiers* 1994).

Though the Trabant now has retro appeal, for much of the 1990s they were symbolic of "backwards" East German manufacturing decisions. *Time* magazine called the Trabant the twenty-seventh worst car ever made. Certainly, while it is difficult to argue that the Trabant was ever a "good" car, we can see the value of Duroplast within a specific constellation of actants. Trabants made sense for East Germany's repressive politics, tight economics, and resource-scarce manufacturing situation, but more than that, Duroplast's material properties coordinated well with the needs of the many other actants in East Germany's postwar era. That it did not succeed much beyond the East German context was not a fault of Duroplast itself, but a shift in the populations of actants that generated material persuasion. The rest of this chapter addresses this strange world of innovative design projects, the materials used in those projects, and the role of technical communicators in facilitating increased rhetorical awareness of important nonhuman actants.

This chapter showcases instances when transportation designs are persuasive, that is, when they are successful insofar as they are widely used, praised by their myriad users, and environmentally sustainable, among other factors. I show that there is a relationship between design methodologies and rhetorical conceptions of *techne*, where an explicit conception of *techne* roughly corresponds to "design science," and where a more abstract *techne* fits with "design thinking." When design teams working on innovative projects are broken up into compartmentalized working groups, based on specific kinds of *techne*, they can easily "lose the plot" of the larger design goals. Innovative projects, then, require a mixture of specific knowledge domains and broader methods that assist designers in charting a path through untested waters. Innovative projects need both senses of *techne* and can result in "design by committee" when explicit *techne* methods take over. I also argue that radically open design methods, like the use of wikis or crowdsourced design processes, while seemingly correspondent to more innovative projects, do not necessarily lead to design success. Though these methods employ distributed working groups and flattened hierarchies of control, material designs often require closer interactions between humans and nonhumans. As of now, transportation designs need to be tested by users and immersed in the specific assemblage of actants in which they will be used. Along with the example of the East German Trabant, I explore electric cabs in early twentieth-century Manhattan – successful when they were designed in close coordination with the humans and nonhumans who make up their actor-network, but unsuccessful when designers pushed the design into new contexts. To this end, I focus on the concept of *hyle*, a Greek term for matter in a specific design context, and suggest that technical communicators and the projects they work on can be *hyle*-oriented and user-centered at the same time.

Design science and design thinking

In discussing large, technically complex transportation projects, like the design of the Trabant, we can ask ourselves what roles designers and their methods have to play in the development process. After all, in considering design, we attend to a beginning – a human intervention into the world and the development of an idea, project, or technology not previously existent. Design strategies can be employed to advance an optimistic undertaking, when the methods employed or the outcome desired is has not been previously attempted. Few scholars of technical communication have chosen to explore design, or design thinking, as a method that aligns with rhetorical study, though we often make use of design principles in our teaching and research. We might describe design as the theories and methodologies that support the *knowledge of* materials and the *knowledge how* something can be produced. For more innovative design projects, designers have come to employ "design thinking" as a methodology more suited to exploration, but often use "design science" when the methods of knowledge generation are more well-known.

Both design thinking and design science are methods for invention, development, and production. Design science can be traced back to the 1960s, when visionaries like Buckminster Fuller and Herbert Simon popularized the big implications of revolutionary designs – many of which were based around innovative engineering projects. Design science can be described as "an explicitly organised, rational and wholly systematic approach to design; not just the utilisation of scientific knowledge of artefacts, but design in some sense a scientific activity itself" (Cross 2001, 51). Simon's landmark *The Sciences of the Artificial* (1969) asked universities and research centers to focus more on rigorous methods centered design and less on theoretical nuance, which he saw as abstract and less applicable to the problems that face humanity. For Simon, "[t]he engineer, and more generally the designer, is concerned with how things *ought* to be – how they ought to be in order to *attain goals*, and to *function*" (4). Design science was a movement that wanted to raise design activities as rationalized and rationalizable methods applicable to the study of problems in much the same way that empirical science was seen to be. The method of design science incorporates a largely positivist view of how designers work – the assumption that there are defined and appropriate design methods for specific fields and their accepted processes. Design science methods are especially adept at building from existing knowledge and aligning with larger discipline-specific goals.

Design *thinking* proponents, on the other hand, oppose the suggestion that there is a "science" to design methods and methodology, and instead argue that designers should understand the *techniques* of the artificial, not the *sciences* of the artificial. This movement involves a greater critical awareness of design as a professional practice, with designers commenting upon the methods used in design, instead of design functioning in relation to scientific methods or fields. Donald Schön's *The Reflective Practitioner: How Professionals Think in Action* (1983) became a touchstone for this shift. Schön pointed out that designers engage in fundamentally different practices than the subject-matter experts with whom they work and emphasized designers' specialized knowledge. He argued that designers build knowledge about the process of design through critical self-reflection. For Schön, this reflection allows a designer to "criticize the tacit understandings that have grown up around the repetitive experiences of a specialized practice, and can make new sense of the situations of uncertainty or uniqueness which he may allow himself to practice" (61). This reflective practice was to give way to what Cross has called "designerly ways of knowing" in his article (2001) and book (2006) of the same title. Cross argued that design, as a field, "has its own distinct intellectual culture; its own designerly 'things to know, ways of knowing them, and ways of finding out about them'" ("Design Research" 1999, 7). Design thinking suggests design is an art, not a science, and that designers employ a range of knowledge:

> Some of it is knowledge inherent in the activity of designing, gained through engaging in and reflecting on that activity. Some of it is knowledge inherent in the artefacts of the artificial world (e.g. in their forms and configurations – knowledge that is used in copying from, re-using or varying aspects of existing artefacts), gained through using and reflecting upon the use of those artefacts. Some of it is knowledge inherent in the processes of manufacturing the artefacts, gained through making and reflecting upon the making of those artefacts. And some of each of these forms of knowledge can also be gained through instruction in them.
>
> (Cross 2001, 54)

Instead of an assumption that design can be systematized as a "science," design thinking identifies heuristics and known concepts that all designers can employ, where the practice of design is a strategically applied knowledge of materials and procedures.

The practice of design is not the same as the practice of rhetoric, nor are designers synonymous to rhetoricians, but productive theories and heuristics for design show a marked resemblance to rhetorical approaches to problem solving. Further, critics have historically dismissed design as the style or the functional "dressing" of an object, very similar to how misunderstandings of rhetoric relegate it to persuasive style, and how the practice of technical communication has been labeled as the "user-friendly" articulation of functional objects. That is, critics have argued that the real work of making something function properly is the domain of engineers – technical communicators using rhetorically grounded methods only step in and help users understand how to use the technology or (possibly) influence surface features. Design theorist and rhetorician Richard Buchanan notes this trend, offering his perspective that "the typical view of design as a styling of the appearance of products is a serious misconception of the actual work of the designer. It is comparable to the popular view of rhetoric as the mere styling of verbal expression" (2001, 194). These more negative explanations of design disparage it as simply the *shape* or *expression* of the "functional" aspects of a particular object.

To some degree, these negative evaluations of design and rhetoric exist because of the nature of the arts of design and rhetoric – because both can be "applied" to any number of situations (Buchanan 1995, 24). The fields encompass not only the skills for making something, but also knowledge of materials and insights on production. Much of Buchanan's work has centered on exploring the connections between design and rhetoric, and he argues that a wider understanding of rhetorical design would benefit both fields. In recognizing design as deeply rhetorical, he suggests that "design is the art of shaping arguments about the artificial or human-made world, arguments which may be carried forward in the concrete activities of production in each of these areas, with objective results ultimately judged by individuals, groups, and society" (1995, 46). Both designers and technical communicators have a deep appreciation for methods that can help influence design to better serve users. Technical communication research has addressed how user-centered design improved design outcomes (Sullivan 1989/2009; Johnson, Salvo and Zoetewey 2007; Redish 2010), how usability testing is a deeply rhetorical practice (Johnson 1998; Salvo 2001; McGovern 2005; Hailey 2011), how design processes should be centered on users' needs (known through its variants as usability engineering and cooperative, participatory, and/or contextual design), and how technical communicators can function with design teams as experience architects (Potts 2013). Because technical communicators have such a close relationship with designers (whatever their domain), they may recognize the similarities between the activities involved in design and rhetoric. While it would be a mistake to equate the two (see Johnson and Ramey, 239), we can align them as arts that are both "concerned with *discovery* and *invention, argument* and *planning*, and the purposes or ends that guide the activities of the subordinate arts and crafts" (Buchanan 1995, 31). We can also identify both rhetoric and design as endeavors concerned with what Aristotle called "a reasoned state of capacity to make" (1140a6).

Techne and design

For rhetoricians, *techne* is most generally known as the larger category of productive knowledge under which rhetoric is often classified. Most would refer back to Aristotle's description of rhetoric as a *techne* in his *Rhetoric* and would generally define it as a "craft" or "art."[4] Quite famously, in *Nichomachean Ethics*, Aristotle defines *techne* as "a reasoned state of capacity to make," where that capacity is responsible for bringing new things into the world (1140a4). But, even though our general understanding of *techne* is the larger conceptual category for arts like rhetoric, sculpture, medicine, architecture, design, etc., its definition has been in contention for over two millennia. David Roochnik, in *Of Art and Wisdom: Plato's Understanding of Techne* (1996), identifies eight different complex understandings of pre-Platonic *techne* and Kelly Pender (2011) notes at least five current definitions: 1) as a "how-to" guide or handbook; 2) as a rational ability to effect a useful result; 3) as a means of inventing new social possibilities; 4) as a means of producing resources; and 5) as a non-instrumental mode of bringing-forth (16). In part, Pender's work clarifies some of Roochnick's historical analysis and his larger assertion that there are two main forms of *techne* – a *techne* 1 and a *techne* 2. A *techne* 1 is a precise, determinate, easily recognizable, teachable, universal, and infallible set of techniques used to accomplish something. This *techne*, what I will refer to as an "explicit" *techne*, might be most closely aligned with a set of direct commands in an instruction manual and figuratively aligned with design science. A *techne* 2, what we can call an "abstract" *techne*, is much messier, less certain, and less precise affair.

Following Roochnik and Pender, we should recognize that the art of rhetoric is never comprised of only explicit *techne*. Many believe rhetoric to be teachable, yet would hold off on calling it determinate or infallible. For example, architects recognize that there is a range of determinate steps that need to be taken when pouring a footer for a house, yet when using new materials they operate in a less precise context with less certain methods. Chefs feel comfortable preparing a plain cake batter, but are less certain when using ingredients they have never before attempted. When operating in less-certain zones of activity, knowledge is less determinate. Commenting on the less determinate form of *techne*, Janet Atwill argues that *techne* is never a representational body of knowledge, because it resists identification with a static, normative subject. For her, *techne* is "knowledge as production, not product, and as intervention and articulation rather than representation" (1998, 2). While we have developed generalized knowledges about how to build a car, for example, every unique manufacturing situation requires some additional broadening of our methods. *Techne* is not simply knowledge about *how* to do something, but also knowledge about the *relationships between different actants*, with a focus on a specific goal. Thus, along with a more explicit *techne*, we have an abstract *techne*.

Most design projects have an overall goal, yet no specific set of precise instructions for how to reach that goal. For example, when internal combustion engine vehicles are converted to run on battery power, those retrofitting the mechanicals often run into a range of compatibility issues they could not have imagined when they began. Though these designers had a goal, they did not know the precise range of techniques they would employ to achieve it. Atwill's description of *techne* as marking a domain of human intervention and invention, though not mastery or certainty, neatly summarizes this kind of exploratory design (2). Successful innovative projects make use of an explicit *techne* as well as an abstract *techne* that "intervenes when a boundary or limitation is recognized" and is able to create "a path that both transgresses and refines that boundary" (48). Recognizing that *techne* is more than a set of rules, but includes an abstract process that is necessarily dependent upon specific actors allows designers to employ it in variable-rich contexts. When the boundaries of what is possible are reached, our capacity to develop new solutions is dependent upon the looser, less constrained interventions that help us, as Atwill explains, "create paths in uncharted territories – to help one find one's way in the dark" (68).

Technical communication researchers have explored this version of *techne* before, too, with Ryan Moeller and Ken McAllister's (2002) exploration of a playful *techne* for practitioners and students, Byron Hawk's (2004) previously mentioned articulation of a posthuman "post-*techne*," and Gerald Savage's (2004) connection between technical communicators and the sophist trickster archetype. Technical communicators, therefore, could point design projects to focus on the larger goals of the project and identify the features of an abstract *techne* that can help get them there. Certainly, they can employ communication skills to avoid large divisions between mechanical engineers, electrical engineers, software engineers, interface designers, and so on, each working wholly within their specialized, explicit *techne*. The argument for fewer distinctions between specialized design knowledge domains can be especially powerful when we look at examples of cutting-edge designs where designers maintained the boundaries between specialized fields, even though the innovation of the project called for greater integration and abstraction.

Innovative design

When design teams tackle innovative projects, ones in which the *techne* they need to employ are largely uncertain or the precise end result unknown, they should avoid drawing clear distinctions between team members' various specializations. That is, innovative projects are often not well-served by a Balkanization of individualized *techne* or design sciences. I do not mean to suggest that the knowledge domains of specialized fields are not relevant, useful, or innovative in their own right, but only indicate that honoring these fields does not have to result in their respective isolation. This is especially important to remember when a project's purpose is to create what has not been developed before, when a project is pushing against perceived boundaries. For technology theorist Carl Mitcham, a wide array of professions do design work, but are too often isolated from each other when working on the same projects with the same goals. He contrasts "artisan" design methods with modern design fields to point out that designers do not often manufacture something *while in the process of design*. That is, these designs are worked out well-ahead of time – something that innovative projects cannot often do. Projects where isolated designers engage in "thinking before making" can result in a decontextualized process where engineers are "are called upon to exercise their creativity in what may be termed calculative ingenuity" and where the "same engineer is also increasingly alienated or separated from actual fabrication" (2001, 33). Such thinking-before-making is roughly analogous to an explicit *techne*. For Mitcham, artisan designs have "seldom involved anything like a separate moment of thinking out or planning beforehand, but proceeded as intuitive trial-and-error fabrication, letting oneself be guided by materials and tradition, and even by personal relationships in the community" (30). Partitioning a design process into segmented disciplines constrains designers from knowing the entire project. This segmentation can also develop into a kind of "design by committee," which is much less helpful in projects that focus on the shifting outcomes of design. While history is littered with examples of unsuccessful designs, or designs that were not publicly welcomed and sustained, we can point to a recent example of just how isolated explicit *techne* can result in an unfortunate production.

The Pontiac Aztek was a crossover vehicle sold from 2001 to 2005, though it may be best known as Walter White's car in the television series *Breaking Bad.* Initially shown as a concept car in 1999, the Aztek was intended as a true multi-purpose vehicle (some versions included an attached tent) that appealed to a younger demographic. Initial interest in the Aztek was high, but when the vehicle finally emerged, it was largely regarded as ugly and awkward. It featured square wheel wells, extra vents and headlights, odd angles, and acres of plastic body cladding. More than one critic pointed to the source of the Aztek's problems as General Motors' corporate design practices, faulting their "design by committee" approach to vehicle design. Steve McConnell (2004) referenced the Aztek's design process as an anti-example in his book-length argument against sequential, compartmentalized software design. His point was not that good designs cannot be accomplished in steps, but that innovative concepts need guiding principles and more agile methods of design and project management. The fault in GM's development could have been in their highly grouped design approach – where the coherence of the concept car was lost as different designers, working under different constraints, goals, and assumptions, produced what they saw as the best, most economically viable solutions. The combined result sold in less than half the numbers GM had hoped for, won dozens of "ugliest/worst" car awards, and was described by the former CEO of American Motors as "indicative of the failed product development system that has been nurtured over there [at GM] for so long" (Weisman 2005). In interviews, GM designers revealed that "the Aztek design had been fiddled with, fussed over, cost-shaved and otherwise compromised until the tough, cool-looking concept had been reduced to a bulky, plastic-clad mess. A classic case of losing the plot" (Neil 2007). The Aztek design process likely suffered from an abundance of explicit *techne* methods and not enough of an abstract *techne's* guiding principles.

This compartmentalization among design was forged along with modernity's assumptions about best practices for fields of study. Instead of a complex, somewhat messy process of design, fields of knowledge were often situated in isolation. As Bruno Latour has argued, we can see modernity itself as a regime of style, one that broke down different fields of knowledge and processes of production into structured taxonomies. The moderns, Latour claims, assumed that the knowledge domains in engineering and science were able to achieve *mastery* over a subject or a process. He notes that while modernity supposed that isolated fields can confidently deal with matters of fact, we now recognize how the messy problems of today should give rise to equally messy, integrated methodologies

(2008, 2). As Latour differentiates in "From Realpolitik to Dingpolitik: Or, How to make Things Public" (2005), we can no longer claim that positivistic methods reveal hard facts, but need to become more invested in abstract methodologies. For Latour, the methods of positivist science should be replaced by design-driven activities arranged around "spheres of concern," where networked problems are understood as only approachable through distributed, solution-oriented methodologies. The more we see problems *in their respective networks*, the more matters of fact are turned into matters of concern. For an innovative technical project, that messiness can be pursued through some attention to the larger contexts of the design process, even when those contexts require designers to communicate across increasing distances.

Some of the most innovative ideas for automobile design emerging today are being pursued through distributed, online, global means – and structured with platforms that facilitate communication between people who are not in the same place. For design teams, this can mean using a wiki or version control software to share documents, while for others it may mean a radical democratization of the design documentation. In those projects, the pursuit of innovative design has coincided with an open-source model of information-sharing where schematics, CAD renderings, technical descriptions, bills of materials, testing reports, and other documents are free and available for all to use. The open source model is relatively new in automotive design, because proprietary designs have been historically been driven by competition and the high cost of manufacture. While no open source automotive projects have experienced mass production so far, likely due to the cost of manufacture and governmental licensing, the concept has spawned a number of interesting projects.[5] However, making a project's documentation open without an existing manufacturing network does not guarantee any vehicles will be produced.

In order to identify successful open-source vehicle designs, we might look toward the model of the individual engineer – the stereotypical inventor-with-a-dream who produced a project car and made the designs free and open (what Carl Mitcham might recognize as "artisan design"). Electrical engineer Gary Krysztopik made all of the documents for his awkwardly named ZWheelz project free for anyone to use, intending to share his experience as a vehicle designer. Smaller-scale projects like Krysztopik's are often more contained and their success defined by their openness. However, larger, more experimental transportation projects tend to place a greater emphasis on widespread public use as a measure of success. Ultimately, they hope to foster a strong, dedicated community who will further the design process and lead to a propagation of the concept. Several projects have used online, crowd-sourced work structures, or similar democratic work configurations, in order to further their projects, though these methods may only reinforce existing divisions between specialized fields. One example of a crowdsourced, open-sourced, innovative vehicle design that failed to thrive is OSCar (Open Source Car), the crowd-sourced project for the design of an electric, lightweight vehicle. After a rapid start in 1999, and some media attention in the early 2000s, the design project stalled at version 0.2. Likewise, development on the P2P (Peer-to-Peer) Foundation's Open Source Green Car (OSGC) has also slowed to a halt. Though we cannot precisely diagnose why these particular concepts did have make it to fruition, one lesson we can take is that the move toward an open-sourced, crowd-sourced design does not mean that fields of knowledge will become similarly open. Like any other, an open-sourced project can focus too much on explicit *techne* fields of knowledge at the expense of abstract *techne* goals. Those working on producing electrical schematics, for example, may have difficulty coordinating with those focused on ergonomics. Open-source wikis are no guarantee of overall design success, despite the shining example of *Wikipedia*, and crowdsourcing, despite its various successful applications, may not yet be appropriate for transportation designs.

Open-source design projects can also become designs by committee (as it is often easiest to delegate work through a wiki or related software), but they hold the potential to merge knowledge domains and disciplines, too. In innovative projects, success often depends upon how team members consider the more abstract matters of concern, as opposed to the specific and determinate rules of production. In some radically new transportation designs, we can see where modernity's strong delineations between domains of knowledge are no longer as useful. Because this kind of design process is messy, it often involves just-in-time production and agile project management methods – methods that work best when design teams are smaller, when the outcome is uncertain, or when the product manufacture may need to be checked against larger development goals throughout the design process. Transportation designs like EDAG Engineering's Light Car, OSVehicle's TABBY projects, and Riversimple's hydrogen fuel cell vehicle are all examples of open-source vehicle design that have achieved a much greater scale of success, with OSVehicle parts already available for purchase.[6] These structured design teams can foreground matters of concern by focusing on abstract *techne* design thinking in ways that might be more difficult to emphasize in distributed collectives or crowdsourced designs.

A strong commitment to an exploratory design requires both discipline-specific explicit *techne* and broader, more experimental abstract *techne*. The examples above also highlight a major factor in the success of an open-source material design project. Specifically, open, crowd-sourced designs that intend to develop a production-ready vehicle may find that distributed, virtual teams have difficulty producing complex *material designs*. Innovation in transportation design is less likely to be achieved through an open and crowd-sourced model primarily because designers also need to build, test, configure, and involve users in the material result of design. Interacting with a wide range of actors, human and nonhuman, is an important component, not simply because material interaction trumps innovative ideas or because materials determine whether a project is successful, but because *the materials themselves are important actants in the design*. Design is a process influenced as much by humans as it is by nonhumans.

There are more examples of successful designs that have attended to abstract *techne* processes and grounded design teams in a material context. The automobile company called Local Motors builds small production runs of automobiles by only partially crowdsourcing their concepts, not their full design. Their free-to-join, online design process accepts public submissions, which are subject to "design checkups" by members of the community and Local Motors staff, who review designs' viability, practicality, and safety ("Design Studio" 2013). Once a design is selected, the initial user-turned-designer works with Local Motors staff to develop the entire vehicle in a process they call "Co-Creation." The vehicle is then produced in a small manufacturing facility located in the same geographic area where the vehicle is intended for use. Local Motors' first vehicle, the Rally Fighter, is a high ground-clearance coupe designed for a rough desert landscape, designed and built in Phoenix, Arizona ("How Local Motors" 2010). Different vehicle concepts that become popular in different parts of the country suggest which designs will be built in the next "microfactory." Chosen designs understand the importance of humans and nonhumans and the specific material conditions of manufacture and use. While Local Motors keeps their productions limited to 2,000 examples, their model of design represents a *techne* sensitive to how materials in design can be persuasive.

Hyle

So far, this chapter has explained that innovative technical projects are best approached when strong divisions between research and production fields are unified through a larger commitment to abstract *techne*. I also claimed that transportation designs are more successful when the designers engage material conditions of design, despite the advantages of online tools that allow for distributed, democratized work. However, I want to explore one additional facet of the *techne* of design mentioned above – the kind of material persuasion that nonhuman actants have in design projects. While collaborative, democratically organized technical projects benefit from a material grounding, we can also identify how the material actants are able to persuade. In part, I want to identify this material form of persuasion because technical communicators are often in ideal positions to support the relations between symbolic and material use in technical projects. The concept of "*hyle*" (ὕλη) in particular helps identify how material persuasion happens. Homer, Thucydides, and Xenophon used the term *hyle* to refer to wood that will eventually have a use other than as a tree growing in a forest. For the Greeks, the common reference for *hyle* was to wood that has been or could be cut down and used in a project of some kind (*hyle* is often associated with carpentry).[7] For Judith Butler, in *Bodies that Matter* (1993), *hyle* is "wood or timber out of which various cultural constructions are made, but also a principle of origin, development, and teleology which is at once casual and explanatory" (7). Butler explains that it is "wood that already has been cut from trees, instrumentalized and instrumentalizable, artifactual, on the way to being put to use" (7). Vilém Flusser echoes this definition in *The Shape of Things* (trans. 1999), suggesting, "when Greek philosophers took up the word *hyle*, they were thinking not of wood in general but of the particular wood stored in carpenters' workshops" (22). In general, *hyle* refers to material on the way to fulfilling a symbolic, material use in a technical project, though it would be a misunderstanding to think of *hyle* as just referring to a raw material.

While Aristotle also defined *hyle*, his definition is unique among his contemporaries, and because of this, we cannot assume that his meaning held wide currency. He used *hyle* to refer to "matter," or in some cases, "material" – which is a broader definition of the earlier meaning as wood about to be a carpenter's hands.[8] Aristotle explained that *hyle* is "that which in itself is neither a particular thing nor of a certain quantity nor assigned to any other of the categories by which being is determined" (1029a18–20). The term *hyle* is general, yet refers to material that exists in immediate context. Or, to put it another way, *hyle* is both the *general form of material potential* and *material in the context of a specific technical project* (Peters 1970, 88–90). Effectively, there are two meanings encompassed by the one term. *Hyle* is a more metaphysical term for matter – the "ultimate substratum" of all things – and, as Ivor LecLerc explains in *The Nature of Physical Existence* (orig. 1972), a specific term for matter in its immediate context (1029a23). In the second understanding, *hyle* is not form, but the potential for change or adaptability in a specific project or purpose. LecLerc argues that *hyle* does not "denote any one specific and definite thing, some sort of [elemental] stuff for example, which can be identified as an ultimate constituent in every physical existent. Rather, the concept of *hyle* towards material is a relative one: the *hyle* is always relative to the particular actual thing" (2004, 116). Weirdly, then, what we have in *hyle* is a *generalized specific*. It is a general term for material that can be used in a specific context. In Aristotle's explanation in *Physics*, "matter [*hyle*] is a relative thing – for different forms there is different matter [*hyle*]" (194b9). It is matter that has general principles, but is unique to the specific physical location in which it is being engaged.

From the standpoint of abstract *techne*, we would say that it is part of the designer's job to understand a project's *hyle*. This would involve knowing the general nature of the materials that are to go into the design as well as how design strategies bring out a specific material's qualities – also in a specific design. Technical projects require a nuanced understanding of how specific materials change, alter, and impact the outcome. The materials in these projects persuade, not because they are able to form coherent arguments, but because they produce different meanings as they are invited into specific designs. In these contexts, the materials, the *hyle*, take on new meanings that are not simply those that humans impose, but ones that emerge from the humans and nonhumans involved in the design. Here, materials persuade as their matter impacts design – just as how the wood of a tree becomes the specific wood in a carpenter's workshop, able to be used in a specific project.

When technical communicators work within a project, they are able to identify how specific *hyle* are applicable in energizing a particular assemblage of humans and nonhumans. If particular materials can fulfill that persuasive role within a technical project for some time, then we might also state that they sustain that argument. While it would be inaccurate to suggest that *hyle* itself is persuasive, the identification of *hyle* can be a way for technical communicators to bring attention to material persuasion in a design. Here, it would be the responsibility of designers and technical communicators alike to understand the ways in which specific materials can be used in specific design projects and help communicate those possibilities. In order to comprehend the importance of this, we can look toward an example in which the designers involved in a transportation technology did not thoughtfully consider *hyle* in relation to the humans and nonhumans and how that callousness contributed to the failure of an alternative to internal combustion engine automobility.

Hyle and electric taxi cabs

By the close of the nineteenth century, the Electric Carriage and Wagon Company, later called the Electric Vehicle Company (EVC), was operating electric taxis throughout Manhattan.[9] These cabs were mainly in the "hansom cab" and "brougham" style and looked very much like horse drawn carriages, *sans* horses. They initially featured two 1.5 horsepower Lundell motors provided by the Interior Conduit and Insulation Company and could average around 8 mph. The Electric Carriage and Wagon Company also built a central changing station to recharge and store batteries – workers could switch out the 1,300 lb. batteries in about 75 seconds, using cranes and hydraulic lifts to move charged and drained batteries around (Kirsch 2000, 43; Mom 2004, 81). Their success was equally rapid. They promised to build 100 electric vehicles in 1897, at a time when the total US output of vehicles of any kind was about 50 per year (Kirsch, 41). Some of this early success was due to the slow advancement of internal combustion vehicles and numerous city-wide bans on steam vehicles, but also that designers understood the *hyle* of their electric vehicles.

The *techne* of these electric vehicles was deeply considerate of the constellations of actants that were involved in the design project. In fact, electric automobiles achieved their success primarily because of a close understanding of their many influential actants, including the limitations of electric batteries. Because they were not mass-produced, and because there were not yet significant standardizations in place for automobile manufacture, electric vehicles could be designed to reflect the specific conditions in which they were used – many were redesigned with almost every large vehicle order. For example, between EVC's second- and third-generation cabs, designers repositioned the batteries to allow for easier removal (and to ventilate acid fumes), increased the size of the passenger cabin to accommodate people and goods, strengthened the brakes to better control vehicle weight, and developed stronger, wider tires (Kirsch, 43). We might call this a kind of iterative design, where designers responded to the users' needs and thoughtfully considered the materials that would best fit those needs. These designs were especially sensitive to the immediate material context, too. When snowstorms hit New York, early internal combustion engine (ICE) vehicles were unable to get around the city, but electric cabs could easily move about because of their heavier weight and special tires. The early electric taxi, then, was designed with a deep consideration of the range of applications, needs, and influences of its immediate human and nonhuman actants. Their success was a result of their messy design process, their focus on spheres of concern, and the designers' attention to the impact of material properties in a specific context.

The designers of these cars were concerned with listening to the nonhuman actants and in applying their knowledge to complex environments in the act of making a new electric cab. Further, they had to apply their *knowledge of* and *knowledge how*, their *techne* that was sensitive to innovative purposes and material. They understood the *hyle* of electric taxi cabs. Much of this attention focused on solving issues of design related to roads, tires, batteries, and drivers – a previously nonexistent constellation of actants. Designers had to consider that most roads in major American cities were unpaved, less than 40% in 1880, and that the paving was of widely different sorts, including gravel, brick, granite block, asphalt, and even wooden blocks soaked in creosote oil (McShane 1994, 60). Each surface rattled and shook vehicles in different ways. In order to handle these wildly differing material conditions, designers developed new tire and wheel combinations. For example, the EVC worked with the Diamond Tire Company and created specially made thick-walled tires for their New York City electric cabs. Designers also considered the weight of electric batteries in relation to overall vehicle weight, looking for a perfect ratio between the size of the battery and the weight of the car. Larger batteries would extend a vehicle's range, but also weigh more, which would then decrease range. Yet, despite all of these different sensitivities to *hyle*, the EVC eventually folded in 1912.

The EVC and other electric taxi concerns in the US failed, in large part because they attempted to spread their innovative designs to new locations without due diligence as to the kind of contextual *techne* and *hyle* that initially made them successful. Flush with early success, in 1899 the EVC issued over $100 million in stock, ordered 1,600 more cabs, and moved to expand their business to new markets, including Boston, Newport, Rochester, Chicago, Philadelphia, New Jersey, Washington DC, and Mexico City (Kirsch, 56–57). Though only some of these expansions came to pass, the EVC seemingly forgot that their designs were tied to New York City and the specific *hyle* that existed there. Historian David Kirsch suggests that "[t]he relatively compact and isolating geography of Manhattan was well suited to the capabilities of the electric cabs," but other locations, with other actants, presented new design challenges (75). First, the roads were not good enough in smaller cities or rural areas to support early automobiles. Many vehicles suffered road-related failures, as battery plates cracked, wheels split, or tires burst. Second, the island of Manhattan hosted charging stations that could handle the limited range of the vehicles, whereas other cities had fewer electric stations and less flexibility to create more (Mom, 80). Third, different batteries were suited to different conditions. A battery that functioned well in New York would perform poorly in Washington, because less range was expected in a densely populated city (96).

More design problems arose when electric car companies became entangled in a battle for market supremacy with ICE vehicles, and their distinctly sensitive design practices were applied to different locales and encountered new and different actants. To compete with ICE cabs, the American electric cab, which had, over time, become lighter and switched to pneumatic tires, began to mimic European ICE car designs. As builders incorporated larger batteries to handle greater distances and higher speeds, vehicles became heavier and forced designers to move back to costlier, solid tires and abandon the pneumatic tire and its advantages (Mom, 95). The early success of electric taxis suggests that electric vehicles could have flourished, as long as designers were sensitive to *hyle* and the matters of concern surrounding how and where their electric vehicles were used. But, caught in the rush to capitalize on new markets, designers became inattentive to the initial balance of explicit and abstract *techne* and their earlier focus on the *hyle* of different materials. When designed with a sensitivity to *hyle* as described above, electric vehicles thrived, but when electric vehicle designers tried to stretch their original designs to a new material context and ignored the specific humans and nonhumans that made them successful, electric vehicles suffered.

Complicating user-centered design

The approach to design explained in this chapter could play a significant role in the positions that technical communicators are able to take up in innovative technical projects. In both the opening example concerning East German Trabants and the sphere of success for electric taxis in turn-of-the-century Manhattan, we find that specific material actants impacted the success of the transportation project. For Trabants, design was constrained by a lack of steel and a deeply repressive state, and for electric taxis, design was limited by battery capacity, geography, weather, and the habits of Manhattanites. In both, we can see that designers succeeded when they paid close attention to how actants interacted in the project and failed when either constellations of actants changed or designers neglected them. From the standpoint of the technical communicator and symbolic-material worker, the imperatives of user-centered design can be stretched to include a concern with nonhuman actants. That is, user-centered design does not need to inherently discriminate between humans and nonhumans.[10] Even though humans are often the most important users in transportation designs, as we design transportation for our needs, desires, and safety, we should not always assume that human actants are *always* at the center of a design process. We can allow a user-centered approach to generate an understanding of all potential actants, nonhumans included.

In the excellent *User-Centered Technology: A Rhetorical Theory for Computers and other Mundane Artifacts* (1998), Robert Johnson offers a useful anecdote that underscores his claims about user-centered design. He explains a situation when, during a bout of particularly freezing temperatures, a person begins bringing his car battery inside at night. Though less common now, this practice was once typical, because a warm battery would start faster than one left outside at night. On the way to work, the driver stops for gas, but finds that his car will not start after he fills up. He is stumped as to why, as are the mechanics in the garage, but then an older mechanic pulls up and identifies the problem right away. Johnson is careful not to suggest that this mechanic's ability is simply due to the difference between experts and novices, but claims that the mechanic possesses an advanced *techne*, a tacit knowledge that comes from experiencing similar situations, and *metis*, a cunning intelligence that guides his ability to act quickly and appropriately. Understandably, Johnson focuses on *techne* as a specifically human characteristic that is not necessarily aligned with the initial intentions of designers. Johnson explains, "the user of the technology has, through *techne*, the knowledge to overcome the potentially inevitable consequences brought forth by domineering and determining forces [. . .] for us, these forces may be technology itself" (52). This story helps Johnson argue that users are also producers of knowledge and that designers should center on this user-generated knowledge if they want to have successful, ethically sound outcomes to their design projects.

If we take Johnson's story and imagine the design process of a new battery, one that is better able to cope with extended freezing temperatures (in actuality, these have already been designed), we are able to understand how a *hyle*-oriented model is not necessarily different than a user-centered one. The primary inquiry into a new design comes from a motivating dissatisfaction voiced by users – car drivers – who are annoyed at either difficult-to-start automobiles or tired of bringing their batteries into their living rooms at night. The redesign process would start with these users, as they are the most vocal and persuasive actants in this constellation. Their sense of convenience, their sense of safety, and their own *techne* are all quite important. However, a redesign would also consider external factors, including temperature conditions for a specific area, terminal end design, potential terminal materials, and the larger material context of existing vehicle design. In essence, design practices spawn smaller *hyle*-centered inquiry methods considered *in coordination* with a user-centered model of design. Such a design practice is still user-centered, as the drivers of automobiles, their knowledge, preferences, and safety are still the expressed, primary concerns, but this focus allows technical communicators to see which materials are most applicable and sustainable within a context of use.

Attentiveness to the *hyle* in design is not an attempt to displace human users in user-centered design, only point out that we have other actants to consider, some of them nonhuman, some of them deeply important to a design's success or sustainability. The Trabant vehicle, after all, was made from a material that was surprisingly effective at doing what it was initially intended to do within a very tightly controlled environment. Yet, the same materials that made the design viable in one constellation of actants also made it radically unsuited for another constellation of actants. A *hyle*-oriented, user-centered design could still attend to the specific needs, desires, and concerns of humans, but would not automatically assume that humans are central in any given design project.

In this chapter, I have built upon previous explanations of automobility and the structure of persuasion presented in Chapter 2 with a specific look at design in innovative transportation projects. I connected the arts of design with what are more generally known as the art of making and identified different approaches to design, both design science and design thinking. Design science correlates to what has been articulated by other scholars as an explicit *techne*, while design thinking is more appropriate for innovative technical projects and can also be described as abstract *techne*. Building from these connections, I showed that projects where the outcome is uncertain require a balance of explicit and abstract *techne*; too much of one or the other and the project is likely to suffer. When explicit *techne* takes hold, those involved in the design can become isolated within their respective knowledge domains, leading to situations where a design loses its innovation and coherence. While speculative transportation projects may be tempted to employ collaborative, crowdsourced online tools, those tools may also make the development of material designs more difficult. These kinds of projects need a material grounding at some stage in their development, because the *hyle* of the project needs to be engaged by the designers. After explaining the value of *hyle* as a way to understand material persuasion, I identified how user-centered design can also include *hyle*-oriented thinking, so as to avoid ignoring the persuasive materials in a design. The following chapter moves away from the focus on material design to the site where humans and nonhumans meet in the use of transportation technologies – the location of the interface.

Notes

1 Ford had experimented with a soybean and hemp vehicle and other cars have been made out of leather and cloth, though none of these were sold in any significant numbers (Ford and Gregorie 1942).

2 Though specific numbers are hard to come by, Trabant pollution estimates the car produced about 1,000 g CO_2/km, opposed to about 130 g CO_2/km for the most polluting new vehicles using a traditional internal combustion engine.

3 The Trabant was also part of what East German officials called the *Bewegungssystem* (movement system), where the transportation network fit the desire to restrain citizens' mobility. The Trabant lacked the power or range to regularly travel long distances and functioned as a technological encoding of the government's desire to limit the East German's range of mobility, or, as Eli Ruben explains, "the radius of permissible travel in East Germany was built into the Trabant's engine itself" (2009, 38).

4 The debate whether rhetoric (as a *techne*) is a theoretical, practical, or productive art is an ongoing and complex one. William Grimaldi (1972) and George Kennedy (1980) explain that rhetoric is theoretical knowledge. Richard McKeon (1946) and Carolyn R. Miller (1989) see rhetoric as concerned with practical action. Scholars such as Janet Atwill, Barbara Warnick (1993), Ernesto Grassi (2001), Samuel Ijsseling (2008), and Alan Gross (2008) have made arguments for *techne* as a productive art.

5 In the summer of 2014, Tesla Motors opened all of their patents, so as to allow other electric car manufacturers the ability to gain from Tesla's advances. Tesla CEO Elon Musk added, "Our true competition is not the small trickle of non-Tesla electric cars being produced, but rather the enormous flood of gasoline cars pouring out of the world's factories every day" (Musk 2014).

6 Other open source car projects include the oddly named c,mm,n car (pronounced "common") from the Netherlands, Germany's Plushundert, and Finland's Sakhoautot.

7 The Liddell and Scott *Greek-English Lexicon* defines *hyle* as "forest, woodland," "firewood, fuel," and "the stuff of which a thing is made, material" (1,847).

8 It is worth noting that Aristotle is often quoted on his beliefs about hylomorphism – often understood as a unification of matter (*hyle*) and form (*morphē*). In *On the Soul*, Aristotle argues that the human soul makes a human being alive, and through what we

would call a metaphysical argument, explains how beings have the potential to change (matter) as well as exist in actual states (form).

9 The electric cab phenomenon was broader than just New York, as by 1897 the London Electric Cab Company and the French *Compagnie générale des Voitures à Paris* (CGV) were operating electric cabs in London and Paris, respectively.

10 User-centered design is a larger approach that situates the user, the people who will make use of a technology, as the central concern. Many studies suggest that iterative, user-centered design methods can significantly increase the chances of design success. Aside from Johnson, other scholars committed to user-centered and/or usability testing design include Steve Krug, Jenny Redish, Michael Salvo, Renee Wever, Jasper van Kuijk, Mica Endsley, Susanne Bødker. The methodology and related methods have also spread to other design-related fields.

4 Interfaces

Like the previous chapter, this one begins with an example that showcases concepts explored in greater depth throughout. However, unlike the previous narrative about the Trabant's design process, this one focuses on the moments of interaction between already designed materials and users in order to add to our understanding of persuasive transportation. I trace the development and use of two novel vehicular interfaces – one that was negatively received and another that has been largely successful. Both came from collaboration between the well-known design firm IDEO and the Ford Motor Corporation. The challenges and successes found in this collaboration highlight just how complicated these spaces of interaction are to design.

IDEO interfaces

By the mid-2000s, car manufacturers had realized that a huge range of portable electronic devices were being brought into vehicles and used while drivers were on the road. These devices included cell phones, but also satellite navigation units and portable music players. Bluetooth became a more common feature, stereos gained additional "auxiliary" ports for mp3 players, and steering wheels sprouted extra buttons, but the systems that immediately surrounded the driver had remained relatively unchanged. The Ford Motor Company realized it needed a way to integrate these devices into their vehicles, while at the same time reducing driver distraction and encouraging conscientious driving habits. For Ford's Chief Interior Designer, Gary Braddock, the goal was to create an advanced information environment. As Braddock explained, "[t]he very first direction we got was: "Do for the automotive industry what the mouse did for the PC Industry" (Walters 2010). No small task, then. In trying to revolutionize dashboard interfaces, Ford wanted to increase their functionality, yet keep them familiar in order to avoid confusing drivers and customers. Braddock admitted, somewhat understatedly, "[t]here was already a task at hand [driving] and overlaying this new interface on top of that was going make it so there's really a lot going on" (Walters 2010). Ford contracted with the design firm IDEO, better known for their work on household products than anything particularly automotive.

In envisioning a new suite of interfaces, IDEO and Ford established the Ford FIVE project, which rethought the steering wheel and the center stack to accommodate new devices and new uses. IDEO's goal was to create more than just a single-vehicle interface design for Ford, but a platform that could spread across their entire lineup. To better understand how users might engage with new interfaces, in new locations, they embarked on an extensive series of human factors testing, first with paper and magnetic prototypes (stuck to an actual dashboard) and later with participant interviews and rough hierarchies of an information architecture. From there, they developed basic arduino sets linked to real switchgear, a dashboard with a touch screen in the center column, and a rough prototype of a steering wheel, all connected to large displays of the racing game Gran Turismo. This usability experience (UX) design led to a final concept, which then went through more rounds of iterative design, recalibration, and UX testing as a production prototype was developed by Ford engineers. The interface platform that resulted was MyFord Touch, first announced at the 2010 Consumer Electronics Show and first implemented in the 2011 Ford Edge. The platform made use of three screens, included Wi-Fi, was open to third-party apps, and ran on the second generation of Microsoft's SYNC system (Trimble 2010).

Unfortunately, the initial response to MyFord Touch was not good. When the system debuted, users were no doubt excited to try out some of the many new integration features, yet found the hierarchy of information confusing, the touch screens slow to react, the voice commands easily misunderstood, and the entirety of the experience more distracting than seamless. Ford responded to these complaints with software fixes, ones that vehicle owners could install themselves or have performed at Ford dealers, free of charge. Despite these upgrades, however, the reception of Ford's infotainment system has remained negative. In 2010, before Ford MyTouch, Ford placed fifth among automakers in J.D. Powers & Associates' "Initial New Quality" rankings. In 2011, they ranked twenty-third (Jensen 2011). Likewise, as Ford has continued to implement MyFord Touch in additional vehicles, *Consumer Reports* has continued to downgrade its ratings for those vehicles (Mutchler 2012). It remains unclear whether all of the problems with MyFord Touch have been identified, or if they are necessarily linked to software issues; some drivers simply may not be happy with the layout, hierarchy, and metaphors contained of the interfaces themselves. However, a different project between Ford and IDEO received much more positive feedback and revolutionized vehicle interface design.

In a different project, but also in 2006, Ford contracted with IDEO and design consultants Smart Design to develop a prototype for another new interface, this one much smaller in scale than the MyFord Touch system. The specific goal of this project was to augment hybrid vehicles in Ford's lineup. IDEO conducted interviews and observations with 36 drivers, in order to understand their habits, and created a basic framework for interaction from those interviews. Participants ranged from drivers who had never seen gas–electric hybrids, to those very interested in efficient driving. Steve Bishop, IDEO's global lead of sustainability, claims that the "big finding was that drivers interested in fuel efficiency were playing a game. They want a high score" (IDEO 2009). In response to this evidence, IDEO and Smart Design focused on an interactive visual metaphor that would discreetly help drivers get better mileage. The result was the Ford Smart-Gauge with EcoGuide. Located just behind the steering wheel, the Smart Gauge features two liquid crystal displays (LCDs) on either side of the speedometer that show one of four data screens:

- Inform: shows fuel level and battery charge status
- Enlighten: represents electric vehicle mode indicator and tachometer
- Engage: displays engine output power and battery output power
- Empower: visualizes the power sent to each wheel, engine pull-up threshold and accessory power consumption

(*Ford Motor Company Newsroom* 2008)

In theory, this interface creates a driving situation where "the gauges and the configurations function as tools that better communicate with drivers, giving them the tools to maximize fuel efficiency" (Eisenstein 2009).

The most significant innovation on the Smart Gauge however, was the specific interface and metaphor of "Efficiency Leaves." Efficiency Leaves are graphic representations of green leaves located on the right LCD screen that "grow" in number whenever the vehicle attains a pre-established level of driving efficiency. As Ford explains:

> Short-term fuel efficiency can be displayed in two ways – either as a traditional chart or using an innovative display that shows "growing leaves" and vines on the right side of the cluster. The more efficient a customer is, the more lush and beautiful the leaves and vines, creating a visual reward for the driver's efforts.
>
> (*Media.Ford.com* 2010)

In 2010, Ford unveiled the SmartGauge in its Fusion hybrid model, which was rated at an EPA estimated 41mpg city/36mpg highway, though early reviews suggested that the vehicle was capable of much better mileage. While one *Los Angeles Times* auto critic achieved 52 mpg, there has been no conclusive study showing that a driver engaging with the Efficiency Leaves provides increased gas mileage (Neil 2008). As the saying goes, while your mileage may vary, the metaphor of the Efficiency Leaves seems to provide a compelling engagement and suggests that the interface metaphor offers motivational interactions for some users. However, in the field of technical communication, as in human–computer interaction studies, the concepts of both "interface" and "metaphor" are not often thought of in a way that facilitates such unique interactions.

Building from the previous chapter, which focused primarily on innovative transportation design strategies, this chapter concerns locations where the human driver and the nonhuman automobile meet – at the interface. In focusing on these moments of interrelation, I consider the relationships that we have with our automobiles, not by studying our emotional reactions, but by treating the ways in which we touch, see, and make sense of the nonhuman surroundings of the automobile. In these interactions, we can see how material forms of persuasion take place and how they develop into everyday habits. First, I survey how the fields of rhetoric and technical communication began to focus on interfaces and how the field of human–computer interaction (HCI) studies has considered the value of metaphor in interface. Finding that there is often an unquestioned reliance upon metaphor, or a reaction against a constrained definition of metaphor, I attempt to re-define how metaphors function within interfaces. Drawing from a definition familiar to physicists, I connect interfaces with actor–network theory's conception of technical mediation and suggest that interfaces are where hybrid agency is created. Examples like IDEO's SmartGauge and the Chevrolet Volt's dashboard interface show that while some metaphors require conscious attention in order for them to be successful, most other metaphors should be catachretic, or "dead." Finally, after focusing on a much older Model T interface, I consider the possible roles for technical communicators in developing the design of unique metaphors for innovative transportation interfaces.

Interfaces and metaphors

In technical communication, scholars have primarily addressed the concept of interfaces with an aim to bring them out into the open, as they are notoriously difficult to critique – in part because interfaces are often designed to be seamless, simple, and "invisible." As Jeff Grabill (2003) has explained regarding computer GUIs, "[t]hey seem natural and inevitable to most people. They are often transparent. Students in my classes can't imagine computers being any other way – and most of the time, neither can I" (465–466). In the last 25 years, technical communication scholars have worked to identify how metaphors encourage or discourage certain interactions. They have argued that technical communicators should have a greater role in the production of interfaces, and that different fields of study could gain from seeing interfaces as rhetorical constructions. Patricia Sullivan's "Human-Computer Interaction Perspectives on Word-Processing Issues" (1989/2009) is the first such argument of note. After performing an extensive literature review, Sullivan finds that human–computer interaction (HCI) and technical communication researchers could benefit from interdisciplinary research, claiming that "human-computer interaction needs to critique out of a base of writing theory and writing process as well as out of a base of technological sophistication," at the same time that "teaching and research using word processing need to incorporate the attitudes that look for the best and that critique the features and programs in use" (25). A year later, Charles Velte would make a similar argument about human-factors research (1990), though the majority of technical communication scholarship on interface would shift toward ideological critique.

Building upon Sullivan's work, Cynthia and Richard Selfe analyzed the ideological structures of software interfaces in their now canonical essay "The Politics of the Interface: Power and its Exercise in the Electronic Contact Zone" (1994). Selfe and Selfe examined the interface as a powerful type of ideological boundary that often mirrors hegemonic values and enables already privileged, logocentric actions. They argued that "[w]ithin virtual space represented by these interfaces and elsewhere within computer systems, the values of our culture – ideological, political, economic, educational – are both implicitly and explicitly mapped, constituting a complex set of material relations among culture, technology, and technology users" (485). Under this critical gaze, interfaces are revealed to be the visual and functional representation of the ideological forces present in their design. As Ann Frances Wysocki and Julia Jasken (2004) summarize, revealing the oft-hidden interface by way of ideological critique became the field's dominant approach for much of the 1990s. This research was "all about trying to get us to see that not being attentive to interfaces [. . .] is to overlook how they shape not only what we see on a computer screen but also anyone who sits down to work at that screen" (32). Because technical communication scholars are often interested in writing practices, and most writing practices are mediated through some form of software GUI and its attendant operating system, we came to see interfaces as a kind of text. Joel Haefner, introducing a special issue of *Computers and Composition* on interfaces, explained that "a basic assumption for scholars in the field of interface studies, a metaphor we live by, is that the computer interface functions as a kind of text, presenting and organizing information, persuading us to perceive and do things in a certain manner" (2009, 136). As part of this attention to GUIs, technical communication scholars have addressed the importance of metaphors, mostly insofar as they function as representations of complex already-coded functions. Often this focus borrows assumptions from HCI researchers, who employ metaphors in the development of GUIs.

While most HCI research mentions the significance of metaphors in interface design, only a few scholars define them. Jenifer Tidwell's excellent *Designing Interfaces: Patterns for Effective Interaction Design* (2005) makes only a few references, briefly suggesting that obvious visual metaphors make good design practices, and Redondo, Bravo, and Ortega's *Engineering the User Interface: From Research to Practice* (2009) offers a scant mention of metaphor. In a more complete bibliography of design and metaphor, Alan Blackwell (2006) finds that "professional designers employ [metaphor] as an unproblematic design property" and that "more recent textbooks, if they mention metaphor at all, warn that it should be treated with care" (492–493). The common use of metaphor in interface design is literary – as a trope or tactic that relies upon a rather reductive reading of Aristotle's description, that is, "giving the thing a name that belongs to something else" (1457b–1457b8). Since the early 1980s, the work of George Lakoff and Mark Johnson, in *Metaphors We Live By* (1980), has also been a significant influence on interface design. Their model of conceptual metaphor claims that metaphors mirror bodily experience and structure cognition and perception. This "image-schematic" model explains that abstract concepts are dependent upon mental images in the mind of the user. Because it also suggests that interface designers can exploit users' existing culturally bounded connections in order to develop more helpful GUIs, Lakoff and Johnson's concepts have gained traction among interface designers. At odds with Lakoff and Johnson, John Carroll's research with the "active learning" theory of metaphor (Carroll and Mack 1985; Carroll et al. 1988; Carroll and Rosson 1994; Neale and Carroll 1997) understands a user's engagement with metaphor as a dynamic process where metaphors need not be ideal in order to create new knowledge or successful interactions. Despite this ongoing discussion, a number of designers have come out against its relevance or usefulness in interface development.

Several notable researchers offer straightforward denials that metaphors should stand as a central concept for building interfaces. Ted Nelson (1990), coiner of the term "hypertext," has argued that metaphors are conceptual dead weight and that the search for ideal metaphors simply distracts from other aspects of design. Likewise, Alan Cooper (1995), one of the driving forces behind Visual Basic, criticized the applicability of metaphors in the design process, claiming that while a good metaphor can be useful, designers most often spend too much time hunting false leads in pursuing the ideal metaphor. He notes a number of major criticisms to what he calls the "metaphor approach" to interface design, including: metaphors do not equate to precise symbols, metaphors do not "scale" well, human intuition about metaphors is not perfect, and meanings change over time. Offering a literature review of HCI materials, Clay Spinuzzi (2001) also concludes that metaphor is "currently the dominant visual-rhetorical framework for designing, evaluating, and critiquing graphic user interfaces," but argues they are not helpful in understanding interfaces (40). Specifically, he suggests:

> The utility of metaphor as a visual–rhetorical design framework has diminished dramatically, and continues to erode. This erosion is due to the spread of computer literacy: Today's interfaces, once meant primarily to scaffold users completely unfamiliar with computers, now respond to users with nearly two decades of experience with GUIs.
>
> (41)

To illustrate his viewpoint on metaphor, Spinuzzi examines the modern personal computer GUI from a historical visual rhetoric perspective. He finds that the many metaphors that have gone into this GUI over the years (folders, files, trash cans, etc.) have been deposited in layers and concludes, "in none of these layers are the metaphors entirely indexical or static for experienced users," so those "interface elements cannot be considered to be metaphors" (41; 42).[1] Rather than ditching the concept of metaphor in interface design, we can identify alternative definitions of metaphor more suitable to the task. A metaphor in an interface does not have to be a static, indexical symbolic feature, fixed in time and space. It can be a fluid representational concept, functioning as a rhetorical feature within an interaction.

Refiguring interface

In considering the design of transportation interfaces, and in moving beyond the dismissal of metaphor in interface design or analysis, we need to shake off the assumption that interface always refers to a GUI. In software design, interface has often meant the design of an end-user's relationship to the code, but in transportation design, interfaces function in a much wider range of interaction. Further, the kinds of interfaces drivers engage with complicate common-sense conceptions of agency in human–vehicle interactions. Fortunately, technical communication scholars have recently produced work on interfaces that counters more reductive definitions. Teena Carnegie (2009) has argued that interfaces are the means and place of interaction, where "the interface functions rhetorically by creating interactivity. In other words, the modes of interactivity are the rhetorical modes of the interface" (166). Carnegie defines and follows these kinds of rhetorical interactivity, which she calls "multi-directionality," "manipulability," and "presence," in order to show that interfaces can be largely invisible, yet deeply powerful. Ultimately, she connects interfaces not with argumentation, but the Roman "exordium," the preliminary or introductory material for an argument that helps to prepare an audience to receive that argument. Collin Brooke (2009) likewise sees interfaces as more than textual or ideological instantiations of GUIs. He defines the interface not as "an object in the same way that more traditional media are [but] as a dialectical space, in [Kenneth] Burke's terms, and a rhetorical space par excellence" (xiii). Brooke, and others, shift the discussion from the idea that interfaces are static, coded representations, to the notion that interfaces are the actants in an ecology of production.

Liza Potts and Angela Harrison's work on the interfaces of message board sites reddit and 4chan during the Boston Marathon bombing (2013) adds to the shift toward seeing interfaces as rhetorical, relational spaces of interaction. They illustrate how interfaces are rhetorical in that their design shapes and is shaped by the actions of users. Potts and Harrison identify four rhetorical characteristics of interfaces that encourage the sharing of information and express concern over how interface design can both impact and emerge from community interaction. With his work on Microsoft Kinekt, David Rieder (2013) has pushed technical communication's definition of interface even further. Rieder identifies how assumptions change when the GUI becomes altered into a natural user interface (NUI) where human bodies and their gestures *become the location* of the interface. A NUI uses cameras in order to translate human bodily positioning, gestures, and movements into commands for the software or operating system. Rieder claims that the shift from GUI to NUI technologies "has transformed what a computing environment is. Today, computing environments are increasingly immersive experiences in which it is difficult to distinguish between a user and a system, which means that the ways in which we achieve a critical and reflective stance is changing" (2013). A location like the Kinekt NUI troubles traditional interfaces and traditional understandings of how metaphors function. Certainly, new developments in transportation interface design will continue to push against the older definitions, especially as NUIs proliferate.

New, gesture-based interfaces (or what are often called "non-contact driver interfaces") for vehicle controls simply do not fit the assumptions about indexical metaphors in GUIs. As an example, we can look at recent, innovative interface designs. In October of 2013, Google purchased the tech startup Flutter, which had been engaged in the development of gesture-based software for home computers. Flutter's software worked with a computer's camera to recognize a small menu of hand gestures that could in turn be used in lieu of mouse- or keyboard-based controls. News of the purchase became more meaningful when, a day later, Google released its patent filing for gesture-based automobile controls. Patent number US20130261871, filed in April of 2012, shows a system where a depth camera and a laser scanner process hand gestures from the driver and synch them to control various vehicle functions (see Fig. 4.1). For example, the "swipe up" gesture made in the region of the air conditioning vent increases fan speed, while a "swipe down" decreases the speed. We might surmise that Google's driverless car could employ some of these same gestures. In such an environment, as Rieder indicates about Kinekt, the human driver performs as part of the interface. The sense of a digitally displayed GUI is largely absent and notions of where agency lies as a human interacts with a nonhuman interface are clouded. Further, as the human driver and host of nonhuman mechanicals, sensors, and cameras must work together, as part of the interface, it becomes easier to see how agency for movement is shared by numerous nonhumans.

As the Google patent suggests, gesture-metaphors and gesture-based interfaces complicate the notion of interface-as-GUI that technical communication researchers have historically assumed. A broader definition of interface, one not reliant upon positivist conceptions of metaphor, would help identify the wide array of persuasive interfaces that we engage on a daily basis. This definition would recognize that interfaces are often defined not by what materials they are made out

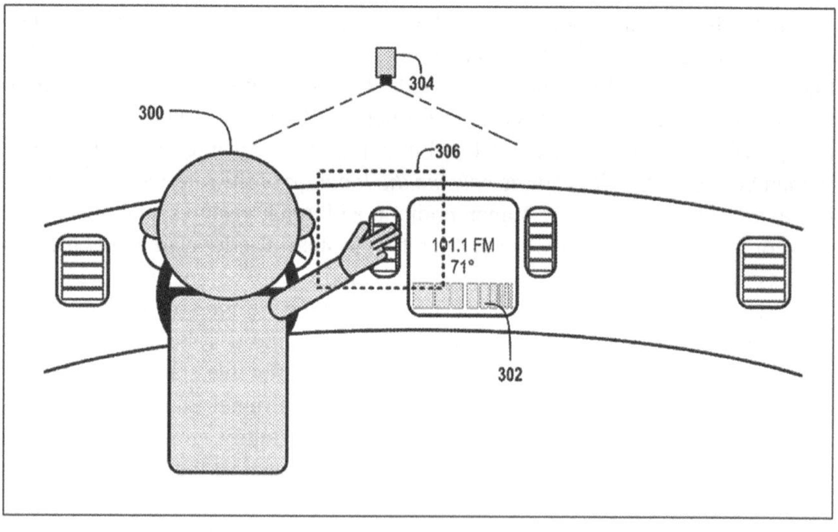

Figure 4.1 Representation of an NUI driver interface from Google's gesture control patent

Source: Hobbs and Chi 2013

of or displayed upon, but by where and how they function and what movements they enable. The *Oxford English Dictionary* gives us several possible definitions to work from. Interfaces can be: an "apparatus designed to connect two scientific instruments, devices, etc., so that they can be operated jointly"; a "place of interaction between two systems, organizations, etc."; or a "surface lying between two portions of matter or space, and forming their common boundary" ("interface" n.d.). In HCI studies, an interface is often considered the software and/or hardware *apparatus* that computer users will need to learn and computer programmers design and code. In Spinuzzi's genre ecology, an interface seems to function as a *place of mediation* between users and their respective activity systems. In physics, an interface is a *surface*, though a special kind, "not a geometric surface but a thin layer that has properties differing from those of the bulk material on either side" ("interface" n.d.). For example, much of our understanding of the surface tension of water involves an analysis of the interface between water and air. While no definition is perfect, I find physics' definition of "surface" most applicable, as "apparatus" suggests that an interface is itself a thing, much like a GUI, and "place" may over-emphasize location.

Defining interfaces as surfaces, however, does not necessarily mean that they take up physical space, nor be identified as a designed technology, like a GUI. Instead, a definition of "surface" emphasizes the multiple kinds of interaction that occur as different actants make sense of each other. An interface is where humans and nonhumans meet, whether through physical contact or symbolic relationships. There exist definitions from technical communication research that align with such a model, as Wysocki and Jasken's 2004 research offered a definition pulled from the *ComputerUser* dictionary that defined an interface as "[a] shared boundary where two or more systems meet; or the means by which communication is achieved at this boundary" (*ComputerUser*). This definition aligns with material physics' understanding of interface, as a thin, non-geometric layer, while allowing us to recognize that both human and nonhuman actants interact at the level of the interface. Granted, if we expect to find a thin, non-geometric layer existing between actants, we might have difficulty identifying one. I know I would. However, this definition is broad enough to allow us to identify different interfaces that provide for all manner of interaction. Within this more expansive definition, we still need to identify how *metaphors* function as smaller elements of a persuasive interface.

In coming back to the problematic understanding of metaphors, we find that rhetorical conceptions of metaphor typically define it as a trope or figure of speech that, in the least developed understandings, works only to embellish language, or in the most developed, functions as a complex communicative act. In either case, nearly all theoretical analyses of metaphor begin by referencing Aristotle's work[2] and his definition in *Poetics*, that "[m]etaphor consists in giving the thing a name that belongs to something else; the transference being either from genus to species, or from species to genus, or from species to species, or on grounds of analogy" (1457b–1457b10). Aristotle's stance on metaphor is frequently misunderstood as 1) one word (and its related concept) *taking the place of* another word, or 2) one word *adding information to* another word. To be fair, there is evidence that Aristotle regarded metaphor as simply a figure of speech, however his definition is certainly more complicated than it first appears, because of the lack of clarity of several key terms in his definition.

In Greek, the word "metaphor" breaks down into *meta* (μετά) and *pheiein* (πηειειν), meaning "change" and "to bear" or "to carry," respectively. John Kirby (1997) suggests that "it may also mean 'across,' so that *metaphora* (μεταπηορα) is literally a 'carrying across' or from one point to another" (532). Another key word in Aristotle's definition is more problematic to translate, however. Specifically, the Greek for "giving," *epiphora* (επιπηορα), is a bit of a puzzle when used in a definition for metaphor, to some degree because both *epiphora* and metaphor share the same root. *Pher* (πηερ) and *phor* (πηορ) translate as "to carry" or "to bring," which suggests that any metaphor would involve a kind of movement, but translations of *epiphora* also suggest more about the nature of this relationship.[3] Though Kirby claims that "we might translate it 'additional assignment,'" he makes reference to the multiple translations that appear in published versions of Aristotle's *Poetics*: "giving," "movement," "transfer(ence)," and "application" (532–533). Each of these possible translations suggests a unique form of interaction and offers us ways to read the relationship between the two things (*onama*). But instead of choosing only one particular translation, it may be more accurate to recognize that *epiphora* is intentionally vague – a notion that seems more than plausible given modern scholars' inconsistency in translating the concept. Noting that *epiphora* is a form of nonspecific *mediation* also moves us further away from the reading of metaphor that assumes it is a *substitution* of one term for another, as *epiphora* cannot be fixed as simply "replacement." Instead, the word *epiphora* describes the interaction that occurs with a metaphor, though the specific nature of this relation (i.e., addition, replacement, transference) is unclear.

It is worth asking, "what is the payoff for redefining both interface and metaphor?" Surely, the definitional backflips of the previous paragraphs need to do something useful for technical communicators trying to make sense of how metaphors can be employed in design. I sort through the language above in part because interfaces are key meeting places where productive action occurs for drivers of vehicles. That is, the way that humans understand the nonhuman actants in a car occurs through the surfaces of interaction (interfaces) and the way that meaning is generated through that interaction (metaphors). As key personnel that understand how the symbolic becomes the material, technical communicators can play an important role in facilitating the production of meaning in transportation projects. This movement between the symbolic and the material is thus useful to understand.

Technical mediation and metaphor

In interface metaphors, different actants encounter each other and generate *something* – not simply by replacing one concept for another, but engaging in a productive interrelation. By reading Aristotle's definition of metaphor as intentionally not specific to the relations that occur at the level of the interface, we allow for a range of non-indexical, non-literal interactions. Even though some metaphors may be subtle and of little value to reflect upon, others are deeply significant (like the Ford SmartGauge with EcoGuide) to how a vehicle's nonhumans function along with its human driver to achieve other goals (greater fuel efficiency). So, beyond the more literal readings of metaphor as static representations, often tied to conceptions of interface-as-GUI, there exists a broader conception of interface as the boundary where the symbolic and the material meet. Technical communicators will be familiar with this interaction, and its design, as the result of "technical mediation."

Latour's concept of "technical mediation" helps us to understand the meeting points between humans and nonhumans and how they are designed into technical projects. Latour identifies technical mediation as a form of "delegation that allows us to mobilize, during interactions, moves made elsewhere, earlier, by other actants" (1994, 52). Any technical mediation is the sedimented result of many other actions, by many other actants, represented in a surface with which actants engage. Key here is that technical mediations function as "delegates or translators of the performance of human actors," because they can "stand in" for the intentions of human actants in various scenarios. Technical mediations are where actants gain strength, not only by simple addition, but through connections with other actants, represented in the specific delegates of interfaces and their respective metaphors. This mediation is made manifest at the level of the interface, where symbolic meaning and intention meets material form. Latour's oft-quoted passage about speed bumps in pedestrian-busy roads helps put this claim into relief.

Latour sees the installation of a mediating speed bump interface as the result of a technique, on the part of engineers, to slow down traffic in order to protect pedestrians. This intention becomes interpreted by drivers as possible suspension damage if the bump is driven over too quickly. That the speed bump is a physical impediment and not a road sign is a delegation of the message from symbolic-alphanumeric intent to material form. Latour uses this example to illustrate the very moment when the symbolic and the material meet:

> Not only has one meaning, in our example, been displaced into another, but an action (the enforcement of the speed law) has been translated into another kind of expression. The engineers' program is inscribed in concrete and, in considering this shift, we quit the relative comfort of linguistic metaphor and enter unknown territory. We have not abandoned meaningful human relations and abruptly entered a world of brute material relations – although this might be the impression of drivers, used to dealing with negotiable signs, now confronted by nonnegotiable speed bumps. The shift is not from discourse to matter because, for the engineers, the speed bump is one meaningful articulation within a gamut of possibilities among which they choose as freely as one chooses vocabulary in a language. Thus, we remain in meaning but no longer in discourse: yet we do not reside among mere objects. Where are we?
>
> (Latour 1999, 187)

Though the metaphor of the speed bump is not terribly difficult to understand, the example emphasizes that interactivity occurs through technical mediation – mediation that is manifest when the human driver, the nonhuman automobile, and the nonhuman road surface, work together at the site of the interface. The speed bump, a technical mediation of the desires of traffic engineers and public safety concerns, functions through metaphor. The driver understands that her car should slow down, the vehicle communicates to the driver that the speed bump should be taken slowly, and the driver, through the vehicle, understands the speed bump as an obstacle to driving quickly and thus as a message to slow down. There are a range of physical and material changes that result, too, from the movement of the driver, to the pressure in the brake line and the grip of the tires on the road surface. In the symbolic-material encounter, the technical mediation employs a visual metaphor of the speed bump, a material interaction, and its myriad interrelations to persuade the driver. This, then, is where the symbolic and the material meet.

Interfaces are not just the boundaries where technical mediations are made manifest and where symbolic and material interaction occur, they are also where agency is produced and shared. As Latour explains at length, a citizen with a gun is a different kind of actant with different agency than one without (see Chapter 2). Such shared agency at the level of the interface (between human and gun) also makes interfaces more difficult to see, understand, and study (as Jeff Grabill explained above). The result of this shared agency is a hybrid, where subjects and objects conjoin with other actants, examples of which, like the citizen-gun and the driver–car, proliferate in everyday life. Much like interfaces, these hybrids are difficult to see because they become "black-boxed" and hidden behind our more vocal claims of control and mastery. That is, we are often said to be in complete control of our cars, thus rendering the hybrid driver–car relation less visible and the interfaces that connect them similarly difficult to discern. Yet, a driver and an automobile cannot be cleaved apart so as to render one or the other in complete agentive control when in motion. These actants are hybrids, formed at the level of the interface – again, the boundary where the symbolic and the material meet.

For a modern example of how an interface and its metaphor, even a surprising one, help to generate hybrid agency, we need not look far. The much-lauded Chevrolet Volt electric car features several metaphors on its digital dashboard interface that function with and persuade the driver to make the most efficient use of the fuel-saving drivetrain. Specifically, the vehicle's Configurable Cluster Display, when in the "Enhanced Mode," shows a rotating green sphere, comprised of leaves, sitting in the center of a vertical gauge. The Volt's promotional materials suggest that when in motion, the driver–car should attempt to "keep the sphere balanced in the middle. If the ball turns yellow and travels above the center of the gauge, the vehicle is accelerating too aggressively to achieve maximum efficiency. If it travels below the center of the gauge, you're breaking too aggressively to capture energy from regenerative breaking" (Chevrolet 2010). The metaphor the rotating green leaf ball serves to enable is something like: "balanced, green leaves equal more efficient driving." Aside from sounding like Chomskian nonsensical, the interactive leaf ball traffics on the assumption that both the color green and images of plant life are enthymematic for environmental responsibility. Yet, even aside from the ecological connotations, the leaf ball represents the potential for different driving habits. Here, green lights can signify a range of more ecologically sensitive driving habits. Though we know that spinning and moving green leaf balls and ecological driving habits are not the same thing, the metaphor often carries through to users.

In this example, we recognize a few things about the nature of unique metaphors and interfaces. Primarily, we find that the metaphor, though not indexical or static, can still work as a persuasive force for the driver–car hybrid. We also find that the interface, which in this case is a GUI, functions like a surface, or the location where visual symbols and objects meet. When a driver reacts to and functions with the symbolic green ball of leaves, new meaning is produced and a new hybrid agent, or an altered version of an existing hybrid, is created. The metaphor of the Volt's rotating green sphere – admittedly a unique, attention-grabbing one – plays a strong role in the interaction with the human driver. Neither human nor nonhuman movement would occur in quite the same way without it. The specific colors of the spinning ball of leaves, or the leaves themselves for that matter, are less important than the technical mediation taking place. How the metaphor functions has to do with the nature of how one actant (the human driver) understands another actant (the Chevrolet Volt) and the technical mediation (the designers' goals) through metaphor (the symbolic spinning ball of leaves) at the layer of the interface (the dashboard GUI). The result, engineers hope, is a complex interactivity that results in more ecological driving habits. All of these different features play a role in how the driver and the car interact and move throughout the world. That said, too many eye-catching and engaging metaphors need to be used selectively, as drivers can only attend to so many.[4]

Dead metaphors

Metaphors can show their remarkable persuasive power, perhaps ironically, through our lack of attention to them – a situation that is completely understandable, because it is impossible to remain cognizant of the constant barrage of interfaces we encounter every day. When an interface interaction has become subconscious habit and no longer provokes active attention, the metaphor at the heart of the interface can be described as *catachretic*, commonly understood as a "dead" metaphor. While Richard Lanham defines catachresis as either an implied metaphor or particularly farfetched metaphor, it is also frequently understood as a metaphor that has stopped being effective in announcing its presence (1991, 31). These so-called "dead" metaphors still influence movement, perhaps in an even more powerful way than metaphors that we pay actively attend to, yet they function at the level of habit and can be considered akin to common sense. Richard Rorty explains that catachretic metaphors:

> are the skeletons which remain after the capacity to arouse the senses – to cause tingles – has been rubbed off by familiarity and long usage. After

> the scales are rubbed off a butterfly's wing, you have transparency, but not beauty – formal structure without sensuous content. Once the freshness wears off the metaphor, you have plain, literal, transparent language – the sort of language which is ascribed not to any particular person but to 'common sense' or 'reason' or 'intuition.'
>
> (1989, 152)

As this chapter has argued, metaphors function as more than linguistic representations, but as elements within interfaces that help facilitate communication between humans and nonhumans. When metaphors become ignored, and the movements that stem from those interactions labeled "common sense," they fall into habit. Yet, what could be more persuasive than a metaphor we no longer notice? There is great power when we quickly, accurately, and habitually perform complex movements in the world. When the act of driving an automobile becomes "common sense," it becomes the default manner of our movement and a powerful lens through which we engage the world. It also becomes a performance of the larger regime of mobility in which we participate.

Catachretic metaphors are those that facilitate movement at the level of preconscious or anteconscious intention – persuasion at the level where human intentionality does not play a role. Most metaphors involved in transportation interfaces operate as what Nigel Thrift has called a "technological anteconscious," where we have offloaded cognition to our designed environment and where we interact with that environment mainly through catachretic metaphors (2008, 10). For example, few experienced automobile drivers would question the validity of a steering wheel interface, because to do so would render the metaphor (that the rotation of a dashboard-mounted wheel is a vehicle's directional capacity) difficult to understand and complicate the ease of movement. Because that metaphor now functions as an anteconscious activity in order for humans to move around in automobiles, we are less able to recognize it as a metaphor and more likely to think of it as a "common sense." ("Of course cars need steering wheels! What else would they have?") In much the same way that Diane Davis' and Thomas Rickert's work explores what unintentional rhetoric looks like (see Chapter 2), the catachretic metaphors we experience every day are similarly persuasive and powerful.

Of course, as users of complex technologies, we depend upon a wide array of catachretic metaphors in the interfaces we encounter, especially if the interfaces require us to build systems of knowledge and habit in order to use them. The vast majority of the metaphors we encounter in everyday life have not only become catachretic, but need to stay that way, at least temporarily, if designers hope to change behavior through a few unique metaphors. A metaphor that technical communicators hope to bring to the conscious attention of users should stand out, not only from other experiences in everyday life, but from the many other metaphors functioning in an interface. While I am not arguing for a particular ratio of "alive" to "dead" metaphors, nor identifying how many unique metaphors a driver is able to attend to inside an automobile, as those questions would require very specific forms of research, I want to emphasize that designing interfaces for human–vehicular interaction requires us to understand how persuasion happens – and that most successful forms of persuasion are both unconscious and habitual. Further, these relationships have existed throughout automobility, as humans have always engaged alive and dead metaphors in the interfaces throughout the automobile. While the metaphors in a vehicle like an older Ford Model T might be wildly new to us now, they would be catachretic to those accustomed to its daily use.

Model T hand brakes

Examples that showcase the claims I have made in this chapter, namely, that interfaces are best defined as a surface where meaning is made and that non-indexical metaphors function in these interfaces, are not limited to modern vehicles. Today, a modern car equipped with an automatic transmission requires the driver to push a button or twist a key to perform ignition and allow the battery to supply electricity to the starter. The car can then be placed in "drive" where the amount and duration of force applied to the accelerator pedal becomes the means for controlling speed. So ingrained is this mode of starting and propelling an automobile, that most drivers experience it as a black-boxed interface and catachretic metaphor. Of course, I am homogenizing an experience here for the sake of an example, as many people experience mechanical issues, disability accommodations, quirky designs, or other interactions that ask them to be more aware of the tasks they perform. New drivers, too, are unlikely to ignore interface metaphors experienced drivers discount. Similarly, older vehicles have different and varied interfaces, which, over time, have congealed into the rather familiar one we have today, so that almost any driver can enter almost any vehicle, and with a quick look around, understand how to engage with the machine. Such is the power of catachretic metaphors. However, when we encounter a vehicle like a Ford Model T, we find remarkably different interfaces for automobile propulsion. Ford Model Ts were the first assembly-line vehicle produced in large numbers – 15 million between 1908 and 1927, when by some accounts, they accounted for half the vehicles in the world (*The Henry Ford* n.d.). A Model T equipped with a magneto presents a completely different interface from a vehicle produced in the twenty-first century.

To start a Model T when the engine is cold, you must first check to see if there is enough fuel in the tank, using a measuring stick, accordingly marked. The tank should be at least half full, and if you need to add gasoline, you will to use a strainer to filter out water and foreign matter. In order to start the engine, turn the key to "off," retard the spark lever by three or four notches (toward you), open the throttle by five or six notches (toward you), put the emergency brake on as far as it will go (in order to release the clutch and apply the emergency brakes), and turn the key all the way to the left, toward "magneto" (see Fig 4.2). Now walk around to the front of the car and make sure the hand crank is pushed in "until you feel the ratchet on the crank engage with the pin passing through the crank shaft" (Pagé, 139). Using your left hand, pull the crank handle with an upward swing, taking caution that the crank does not spin around and break your wrist (see

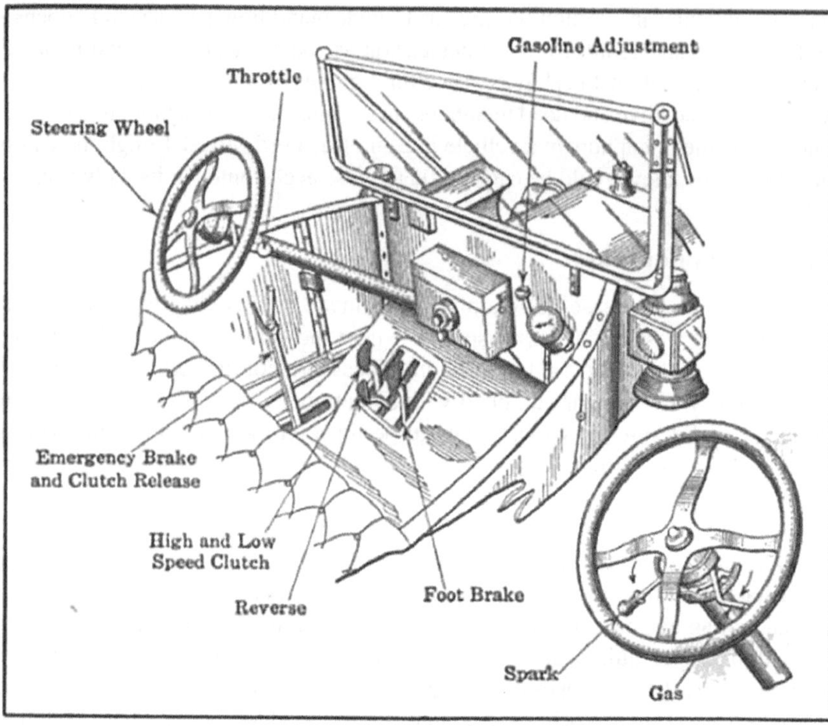

Figure 4.2 Depiction of the driver's interfaces in a Ford Model T
Source: Pagé 1915, 143

Fig. 4.3). If the weather is cold or the car has not been driven recently, you will also need to pull a small choke wire protruding from the lower left corner of the radiator while turning the hand crank a few quarter turns. You can now walk back around to the inside of the car, adjust the spark advance halfway down the quadrant and let the motor run so that it may heat up. In order to move the Model T, release the parking brake by placing it either in the middle position (to enable neutral and first gear), or all the way to the floor (to enable neutral, first, and second gear), press the brake pedal, and slowly press and hold the clutch pedal all the way to the floor to propel the vehicle forward, while slowly advancing the throttle. In order to move to high gear, release the clutch pedal halfway, reduce the throttle by moving the lever forward, push the parking brake to the floor, fully release the clutch pedal, and re-adjust the throttle and spark advance if necessary. Braking, reversing, and parking a Model T are steps best left for another time.

To the eyes of someone familiar with modern vehicles, a huge array of interfaces stand out in the Model T interface, including the use of a separate pedal to engage reverse, due the nature of the Model T's "planetary" transmission, and the necessity of a steering column-mounted lever that provides electricity from the magneto to alight the gasoline mixture. Likewise, the largest metaphor for

Figure 4.3 Turning the hand crank on a Model T

Source: Pagé 1915, 138

starting the Model T might be the somewhat dangerous hand-powered crank at the front of the vehicle. Drivers might understand this hand crank as a metaphor for "helping" the crankshaft to turn, so that the spark may reach the fuel and propel the crankshaft further. No doubt that the experience of driving a Model T would be somewhat bewildering, because of the high number of new metaphors in new interface engagements. These metaphors influence how we consider the car as a mode of transportation and alter our perception of what kinds of movements are possible. Yet, at the same time, we would also notice some metaphors that are largely familiar to us – for example, that of the steering wheel. Though the starting procedure is certainly daunting, the steering wheel would be recognizable and, after acclimating to the driving position and the turning ratio, it would be once again nearly invisible to the user.

For a driver in the early 1920s, when Model T's had achieved worldwide dominance in the automobile market, most of the metaphors would have been catachretic. However, if we look at one specific metaphor, one that would be unfamiliar to today's drivers, we might be able to hypothesize as to the kinds of *kinesis* movements that a specific catachretic metaphor can produce. I would argue that the dominant metaphor for propulsion in a Model T lies in the use of the hand brake, not the gas pedal or the clutch (more familiar to modern drivers). The hand brake offers the possibility of different gears, and thus the availability of speed. The driver only interacts with the clutch in low gear and the throttle lever only as a smaller level adjustment. The hand brake, however, establishes the possibility of speed. Movement forward is based on the knowledge of which of the two gears are needed, first or second. First gear requires constant pressure/attention, and second gear requires little physical exertion (similar to the metaphor of a cruise control). Thus, the basic metaphor for propulsion in a Model T is: "the position of the parking brake equals the potential for speed," which is different from a modern car with an automatic gearbox, whose basic metaphor is something like: "force, degree, and duration applied to a lever (accelerator) equals propulsion."[5] This difference may appear insignificant, for what does it matter that a hand brake is understood as the cause of acceleration rather than a floor-mounted gas pedal? This hand brake, which both works in coordination with the clutch to determine gear, produces an assumption for the human driver: forward propulsion is a result of hand brake adjustment. Functioning with this interpretation of metaphor results in movement, namely, propulsion as two distinct kinds of speed (first gear at 15 mph or second gear at 30 mph). The forms of *kinesis* motion that resulted for those naturalized Model T metaphors, then, might be something similar to seeing automobile mobility as either happening at either a low speed for city driving or a cruising speed for country touring. It is possible that the Model T could have been seen as a storehouse of speed, held back by the parking brake. While these hypotheses are more general, the goal of such hypothetical work is not to produce accurate version of 1920s driver–car understandings, but to show that human–nonhuman persuasion is generated through metaphor at the boundary of the interface.

Interfaces and technical communication

Despite the sometimes metaphysical tone of the above discussion on the role of metaphor in interfaces, we can see meaningful connections to how technical communicators can help design teams focus on metaphor in the larger goal of promoting sustainable interactions. Specifically, useful metaphors help drivers add meaning to a context, aggregating information designed into technical mediations. These technical mediations can lead to new *kinesis* movements and eventually form into habits. In turn, these habits function as rather powerful forms of persuasion, because they do not attract the same kind of conscious attention as more unique metaphors. User time and attention are limited during vehicle operation, so new metaphors can only demand a small portion of that attention, lest the user be too confused, overwhelmed, or distracted by the intensity of the experience (see the concept of "most advanced, yet acceptable"). Because the technical communicator is especially sensitive to the users and their needs, the goal becomes to provide the right amount and intensity of new metaphors to provide a meaningful transportation experience and avoid either information overload or unnecessary complication.

Alan Blackwell, after surveying a range of conceptions of metaphor in HCI design practices, concluded that HCI practitioners and researchers commit an oversight in either discounting the role of metaphor or seeing it as a formal, isolated, and simple component to interface design work. He argued that:

> After 10 years studying UI [user interface] metaphor, including analysis of its historical and philosophical origins, as well as contemporary research trends, my own opinion is that we must pay more serious attention to the creative richness and affect that is valued in professional design. Creative metaphor may continue to be a useful generative tool for the designer, but an even more significant change would be to recognize its value as a creative experience for the user.
>
> (2006, 522)

The technical communicator as symbolic-material worker, as a professional committed to articulating the ways in which symbolic communication relates to and results in material design and movement, is a key part of this increased attention to creative interaction. There is room for increased creativity in the design of novel metaphors, as new dashboard interfaces from Ford and Chevrolet have shown, and technical communicators can help articulate the ways in which drivers make sense out of the metaphors they encounter. Such deep considerations of metaphor are only going to become increasingly important, too, as vehicle designers begin to expand the number and variety of semi- or fully-autonomous vehicular abilities.

For example, we can consider one surprisingly evocative metaphor for autonomous vehicle design – the "H" metaphor. This metaphor was initially developed as a way for designers to conceptualize the position and goals of a user when operating a vehicle in uncertain or complex circumstances, especially when a driver's attention is focused on more than the road ahead. Flemisch et al. were the first to articulate the reasoning behind the metaphor in a NASA technical memo in 2003. They describe the H metaphor as follows:

> Imagine you are riding your bicycle through a wooded park. It's a beautiful day outside, but you are late for an appointment so you're in a hurry. There are crowds of people around enjoying themselves and there are also other cyclists riding nearby you. You're trying to avoid hitting anything and also get to your appointment on time, but you're not very familiar with this park and you need to keep referring to your map. The problem is that even though you're a skilled cyclist, it's very difficult to steer your bike and read a map at the same time with all these trees and people around you, so you are constantly forced to stop. As the time for your appointment draws nearer you keep thinking to yourself – could there be a better way to steer through an environment with obstacles without having to stop every time you need to do something else?
>
> (Flemisch et al., 3)

According to the authors, the answer to such a question is, surprisingly, a *Horse*. On a horse you would be able to read a map "and be confident that you would not hit any trees or run into people because horses instinctively avoid obstacles" (7). The horse provides haptic feedback so that the rider can know what the horse is doing, the horse will react to danger and avoid crashes, and the rider–horse hybrid operates at more than one level of attention. The goal of HCI designers working within driver–vehicle environments (DVEs) would not be to create autonomous vehicle interfaces that simulate the horse-riding experience, but to design with the same metaphoric relationship in mind. If the H metaphor catches on for autonomous vehicle design, technical communicators would be able to provide a great deal of assistance in seeking out extensions of this metaphor through the interface design process.

The role of a technical communicator in working alongside interfaces designers in innovative transportation projects could also be to facilitate the development of useful metaphors for users. Keeping in mind that many users will be new to autonomous features, and need a range of comfortable catachretic metaphors to rely on, technical communicators will want to facilitate the gradual acceptance of new metaphors. With a complex understanding of how metaphors are persuasive, technical communicators could ensure that any new interfaces that demand conscious attention should be integrated carefully, with an eye to which habitual actions are there in support of the new ones. Further, the larger goal within any interface design should likewise be the creation of sustained and sustainable human–vehicle relations. By "sustainable," I mean ones that allow the driver–car to perform movements again and again, without undue stress, and ones that promote safe, ecological driving habits. While interfaces will likely always present a challenge to designers, technical communicators, as symbolic-material workers attuned to technical mediations and the symbolic meaning generated through use of those mediations, are able to help facilitate more verdant relationships.

In this chapter, metaphors and interfaces have been shown to be more complex than they are often depicted in scholarship – and transportation projects are likely to continue to push against commonly accepted definitions of interface. I began the chapter with a narrative about the design firm IDEO's work with Ford, both in successful and less-than-unsuccessful projects, showing the complexities of working with human–nonhuman relations in interface designs. While technical communication work on interfaces stems from the late 1980s, much of this research concerned graphical user interfaces and often defined interfaces in terms of GUIs. More recent work has explored the persuasive interface beyond the GUI and borrowed definitions of interfaces from other disciplines. Human–computer interaction scholarship, for example, has provided technical communication with a number of definitions of interface, and how metaphors function in an interface, though not all of these definitions are as helpful as they could be. I have offered an alternate definition of metaphor and interface in order to show how technical mediations are encountered as persuasive relationships between humans and nonhumans. Using the example of the Chevrolet Volt, I showed how some interfaces generate compelling metaphors, but warn against too many novel interactions in a driver–vehicle environment, as drivers heavily rely upon catachretic metaphors, too. Despite the car's age, we can see these same conceptions of metaphor in Model Ts, though technical communication's understanding of metaphor will only become more important with the advance of semi-autonomous vehicular technologies. From here, the next chapter moves on to consider how interfaces enable more complex kinds of movement – trajectories of movement from one location to another.

Notes

1 I should mention that I have no issue with Spinuzzi's larger argument in this article – that genre ecologies within an activity theory framework are useful in understanding interfaces. Activity theory and Spinuzzi's conception of genre ecologies represent extremely productive analytical and design methods. Spinuzzi's larger claims are significant: that the metaphors built into software several decades previous no longer function the same for users, and that complex interactions cannot be fully understood through just examining metaphors. That said, I believe metaphor can still be a useful way of understanding the logics built into interfaces and how users understand those logics.

2 Aristotle was not the first to use or describe metaphor; we can look as far back as the *Epic of Gilgamesh* for the term and into several pre-Socratic philosophers for partial descriptions.

3 Though it may be obvious to some, Aristotle's definition of metaphor is itself a metaphor. Any concept that functions as metaphor does, that is, as a conceptual and meaning-making trope, is necessarily going to have to use metaphor in its own definition, because it has no "literal" equivalent. As Paul Ricoeur explains, "[t]here is no non-metaphorical standpoint from which one could look upon metaphor, and all the other figures for that matter, as if they were a game played before one's eyes" (1981, 19).

4 One concept many designers hold as a commonplace is that of "most advanced, yet acceptable" (MAYA). First coined by designer Raymond Loewy, the phrase refers to a balance between engineering and design advancement and the public's acceptance and use of those advancements. Too many innovative features can turn an innovative design into a confusing mess for users.

5 It is also valuable to note how distinctly gendered and ableist this particular technology is. That is, the force required to keep a Model T going at half speed is significant and assumed a driver with considerable leg and foot strength.

5 Logistics

While the previous chapter zeroed in on the relationships that emerge between humans and nonhumans at the boundary of the interface, this chapter broadens in scope in order to investigate how those interfaces provide trajectories of movement through the world. While some consider these chains of connected interfaces to be the *infrastructure* of modern life, here, I identify them as *logistics*, which are structures and experiences that are only realized when users actively engage in them. This ability to get from point A to point B is one facility that emerges from groupings of human and nonhuman actants. While many in the US are able to plan vehicular journeys around an internal combustion engine car, and are relatively confident in their ability to find gas when necessary, they are often less confident in sorting out the major interactions needed to power, for example, a plug-in electric car or a hydrogen vehicle. Such concerns are best represented in the very public fight that occurred in the winter of 2013, when Tesla Motors asked journalists to test their electric charging network.

Tesla logistics

In 2012, Tesla Motors delivered their first all-original vehicle, the Tesla Model S, an all-electric, plug-in sedan powered by either a 60 or 85 kWh lithium-ion battery pack with an EPA estimated range of either 208 or 265 miles, respectively. In order to facilitate the new Model S rollout, Tesla began installing 480 volt charging stations, known as Tesla Superchargers, along major highways and roads. A 120 kW Supercharger (see Fig. 5.1) can half-charge a Model S in as little as 20 minutes and provide for up to 170 miles of additional driving in 30 minutes (*Tesla Motors*, n.d.). Reviews of the Model S were quite positive, though some journalists felt that the Supercharger network deserved its own evaluation. *The New York Times*' John Broder provided an early review by driving a Model S from suburban Washington, DC to Boston with an overnight stop in Groton, CT. The test went well enough at first, as he stopped in Delaware to charge the car, spending 49 minutes at the charging station and noting an estimated driving range of 242 miles. After that point, however, the test went south as the journey moved north, as the Model S's estimated range dropped faster than the miles passed – in the following 68 miles of driving, the car registered 85 miles of range had been used. Broder

Figure 5.1 A Tesla Model S plugged in to a Tesla Supercharger

Source: Hicks 2013

claims that at this point he began to follow Tesla's range maximization guidelines, including turning the climate control to low (it was winter) and setting the cruise control to 54 miles per hour. After calling Tesla officials about what else to do and pausing in Manhattan, the Model S ordered him to "Recharge Now" just before he found a Supercharger station. After an hour-long charge, the car registered an additional 185 miles of range, at which point Broder, satisfied, drove to a hotel. In the morning, the car registered only 25 miles remaining, and after Broder performed a process meant to "condition" the battery, only 19 remained. Tesla then found an unofficial charging station for Broder at the Norwich, CT public utilities, which partially charged the vehicle. Soon after, on an exit ramp in Branford, CT, the Tesla shut down and coasted to a halt at the side of the road. Retrieved by a tow truck, the car was then fully charged and driven to a Tesla dealership in Manhattan. Calling about the problem, Broder was told by Tesla's chief technology officer, "[i]t takes more planning than a typical gasoline car, no way around it" (Broder 2013a).

The story caused a significant public relations problem for Tesla when it was published in early February, 2013, and Elon Musk, Tesla's CEO, felt it necessary to respond. Writing on to Twitter to suggest that Broder's review was unethical, Musk (2013a) then penned a fairly thorough takedown of Broder's article. He revealed that ever since a particularly fraudulent review of a Tesla Roadster on BBC's *Top Gear* television show, Tesla has kept vehicle data when lending their cars to journalists. The data on Broder's test, according to Musk, showed that what was reported in the article did not match what occurred during the drive. Specifically, Musk identified that while the final leg of the trip was 61 miles,

Broder disconnected the charge cable when the vehicle displayed only 31 miles of power remaining. Tesla also claimed that Broder never set the cruise control to 54 miles per hour, turned the climate control up instead of down, and perhaps most damning:

> When he first reached our Milford, Connecticut Supercharger, having driven the car hard and after taking an unplanned detour through downtown Manhattan to give his brother a ride, the display said "0 miles remaining." Instead of plugging in the car, he drove in circles for over half a mile in a tiny, 100-space parking lot. When the Model S valiantly refused to die, he eventually plugged it in.
>
> (2013b)

Providing annotated charts to support his claims, Musk accused Broder of deliberately submarining the test, constructing what he called "a no-win scenario for any vehicle, electric or gasoline" (2013b).

At this point, Broder responded, arguing that the data did not tell the entire story and that his reporter's notebook showed that he recorded the events as they happened. On one hand, Broder makes the understandable argument that data does not explain his specific context: in the car, on the interstate, trying his best to get from point A to point B while recharging the vehicle along the way. Yet, on the other hand, Broder's claims are odd. His assertion that Tesla told him to "alternately slow down and speed up to take advantage of regenerative braking" makes no sense, unless a seasoned automotive reporter like Broder was somehow able to subvert the second law of thermodynamics (2013b). Eventually the *New York Times* acquiesced and acknowledged that Broder's notes were imprecise. Later, *CNN* and other news outlets replicated the journey with no problems, and public opinion appeared to side with Musk and Tesla.

Though the details are entertaining, this story is as much about the larger issue of "range anxiety" as it is about a specific journalist's methods. "Range anxiety" describes the fear of running out of stored energy with an electric, plug-in vehicle, first used in reference to General Motors' EV1, a car made famous by the movie *Who Killed the Electric Car?* While the concept can be applied to almost any transportation technology, it remains linked to the electric vehicle because its energy storage capacity and network of "refueling" stations are often compared to much more ingrained transportation systems like that of the internal combustion engine and its attendant gasoline stations (though there was "range anxiety" with the first ICE automobiles, too). The term carries so much weight that General Motors attempted to trademark it when they released their Volt plug-in hybrid (Schultz 2010). In order to overcome this concern, companies have been developing electric plug-in-specific GPS software, which plots journeys that correspond with identified charging stations. As of the time of writing, Tesla vehicles include two software applications called Trip Planner and Range Assurance that can plot driving routes along Supercharger stations (Tesla Motors 2015). Similar technology is also being researched by the SINTEF organization (*Stiftelsen for industriell og teknisk forskning*) and involves the automatic calculation of routes and reminders as to where drivers can safely recharge their vehicles (SINTEF 2013). Range anxiety is a concern born from our faculty for logistical thinking. It happens when we question the ability of a logistic to carry us all the way to our intended destination. Likewise, conversations we have about range anxiety showcase our concern over how to coordinate our mobility desires with our ability to calculate transportation connections. Range anxiety is not so much about a specific car or specific technology as it is about how that technology is situated in a larger network.

The ability to drive an electric vehicle from DC to Boston involves a kind of persuasion, though not one that is immediately recognizable. However, when we attempt to understand the journeys that are available to us, as we think in coordination with the mobility technologies at our disposal, we think in terms of delivery: how we can deliver ourselves and how we can be delivered. This chapter aligns recent conceptions of rhetorical delivery with logistics, first by identifying rhetorical delivery strategies for material contexts, and then by briefly exploring the history of logistics. Employing Aristotle's explanation that all people have a faculty for logistics, I describe why logistics is best defined as an act of connecting interfaces and gathering them together. The recent emergence of car-sharing and "transportation network" companies help show that logistics are not persuasive because of any specific technology, but because they set up multiple interfaces and allow users to gather them together while they are in use. I then move toward a much bigger example, that of electric vehicles at the turn of the twentieth century, and show how even though their downfall has been blamed on the limitations of the electric battery, we can also identify how their specific logistic was largely unconvincing. The implications for technical communicators working in transportation projects concern how they can help design more, better, and safer logistics "scripts" for a greater number of people.

Rhetoric and delivery

Until very recently, delivery had been one of the most marginalized aspects of rhetoric. Traditional conceptions often focused on a rhetors' performance in delivering language, in elocution, yet avoided addressing the material conditions or consequences of delivery. Even Aristotle wrote that delivery had not been taken seriously as an area of inquiry until his treatment of it in book three of *Rhetoric*, though his own treatment is relatively modest. Most of his commentary focuses on delivery's connection with poetry, and how a voice could be managed so as to express different emotions or adjusted to achieve rhythms appropriate to different topics and audiences (1403b15–1404a12). Cicero treats delivery with further attention, finding that any argument is only as powerful as its methods of delivery allow, and calling delivery "the one dominant factor in oratory. Without it, even the best orator cannot be of any account at all, while an average speaker equipped with this skill can often outdo the best orators" (2001, 213–216a). Likewise, Quintilian's *Institutes of Oratory* praises the value of delivery while locating it as a technique of bodily comportment and vocal modulation. Most commentaries on delivery from Greek and Roman rhetoricians focus almost exclusively on human-centric elocution and bodily comportment – concepts in rhetoric that dropped away from critical study with the increase in written literacy and the rise of print. As Jim Porter (2009) has noted, "[b]ecause delivery came to be associated almost exclusively with speech situations and with functions of the speaker's body (voice, gestures), it seemed less relevant, if not irrelevant, to written composition than the other canons (particularly *dispositio* and *elocutio*)" (207).

In the last few decades, delivery has been re-evaluated as applicable to different aspects of written communication. Robert Connors (1983) identified how delivery pertains to print documents, though concluded that most writing scenarios had, at the time, very few publishing options to consider. Sullivan (1992) and Dragga (1992), among others, shifted the focus on delivery to multimedia document design because of delivery's applicability to the technical aspects of digital publishing. Faced with a wider range of layout and compositional options in the world of word processing and desktop publishing, technical communication scholars re-established delivery as a valuable part of the composing process. Yet, John Trimbur's (2000) comments about delivery lamented that neglect of the canon resulted in a conflation of the "activity of composing with writing itself" and led rhetoricians to "miss altogether the complex delivery systems through which writing circulates" (189). Carolyn Rude (2004) similarly found that technical communication's relationship to delivery only covered elements such as document design, visual design, and concerns over medium, where delivery was a conclusion to a finite act that neglected reception. In some of this scholarship, delivery becomes equated with "medium," and the extra-textual aspects of written communication are thus folded into considerations of delivery (see McCorkle 2012, 30). Around this same time, researchers began to understand that delivery could apply to a broader range of rhetorical considerations in an age of emergent crowdsourcing, increased amateur production, and the circulation of online content.

Research on delivery in the 2000s was marked by growing attention to circulation, which followed the shift away from print document design to digital document design, and from a general neglect of textual reception and re-use toward an emphasis on circulation and textual economies. DeVoss and Porter (2006) suggested that "a renewal of interest in delivery requires that we take up the question of 'economies of writing'" that emphasize the distributed, material factors involved in persuasion (194). How these economies have been investigated has varied, from DeVoss and Porter's concern with the politics and economics of distribution practices, to Jenny Edbauer's (2005) focus on public rhetoric and the circulation of meaning in complex rhetorical ecologies. For technical communication, the focus has not always explicitly concerned the canon of delivery, but the many ways that communication happens in a distributed workplace and how information is circulated, re-used, and employed. Valuable research has addressed content management (Clark 2008; Pullman and Gu 2008; McCarthy et al. 2011; Anderson 2014), issues of genre in activity systems (for example, Bazerman 1994; Yates and Orlikowski 2002; Spinuzzi 2003, 2004), distributed work (Nardi and Whittaker 2002; Nardi et al. 2002; Slattery 2007; Spinuzzi 2007; Pigg 2014), and single-sourcing and Darwin Information Typing Architecture (DITA) (Albers 2003; Carter 2003; Dayton and Hopper 2010; Swarts 2010; Evia et al. 2013). While we have broadened notions of circulation and delivery to include technologies as well as texts, a more inclusive conception of delivery in technical communication includes both humans and nonhuman actants, the material spaces of those actants, and how they create strategies for movement.

Rhetorical delivery can be applicable to more than physical and oral expression or textual and visual publishing. Delivery can also help explain how people make choices about their trajectories of movement through the world, as they engage with a wide array of interfaces and make decisions about material affordance, coordination, and velocity. Edbauer's work on rhetorical ecologies, in particular, helps frame this direction for delivery. She looks to Ash Amin and Nigel Thrift's (2002) work on cities to explain how urban spaces are more than static locations, but affective, material interactions that function in an ecology. Amin and Thrift explain that "cities exist as a means of movement, as means to engineer encounters through collection, transport, and collation. They produce, thereby, a complex pattern of traces, a threadwork of intensities . . ." (81). Edbauer argues that a city is as much about these patterns, threadworks, and intensities as it is about material boundaries, yet both are important in the everyday experience of understanding and negotiating physical space. Key for Edbauer in this rhetorical ecology is that movement is necessary for understanding circulation. Similar to how Edbauer has worked to bring considerations of movement back into public rhetoric and shift the conversation from a sedentary rhetorical context to a mobile, shifting ecology, this chapter attempts to bring considerations of nonhuman mobility and shared agency back to material delivery.

In order to think of the strategic decision-making processes that go into understanding material delivery, we also need to recognize that delivery, in the case of automobility, is emergent from the contexts in which humans are immersed. Our understanding of delivery in this context *comes from and is generated by* the humans and nonhumans who comprise the significant relations in any trajectory. Movement through space is not "owned" by humans, but emerges as humans are in the process of engaging with a wide array of interfaces and capacities. Jim Ridolfo and Dànielle Nicole DeVoss (2009) have explained this capacity for estimating the nuances of delivery in a similar way as "rhetorical velocity." Rhetorical velocity is "a strategic approach to composing for rhetorical delivery. It is both a way of considering delivery as a rhetorical mode, aligned with an understanding of how texts work as a component of a strategy" ("intro"). Perhaps more importantly, rhetorical velocity refers to an estimation of appropriate speed for information movement, that is, "the understanding and rapidity at which information is crafted, delivered, distributed, recomposed, redelivered, redistributed, etc., across physical and virtual networks and spaces" ("intro"). Though Ridolfo and DeVoss focus on texts and textual re-use in complex information ecologies, rhetorical velocity can also pertain to transportation and non-textual conceptions of delivery. If we replace our assumptions about texts with concerns for movement in the world, both human and nonhuman, we see that rhetorical velocity strategies can help us estimate the movement of people and things. It would be largely similar, I argue, to the human faculty for understanding logistics.

What are logistics?

As a practice, logistics has always been complex and diverse, as it attends to a massive array of potential variables and interacts with different fields in business, engineering, and military strategy. If we look toward ancient logistics, we see large-scale calculations made by specially trained professionals frequently connected to the military. Some of the records we have for these calculations come from historians like Herodotus, who recorded the requirements of Xerxes' army as it invaded Athens in 480 bce.[1] More often, however, ancient logistics were only recorded when they went horribly wrong, as in Arrian of Nicomedia's account of Alexander the Great's march through the Gedrosian desert.[2] Arrian catalogs the result of poor logistics: lack of water for soldiers and animals, no free beasts of burden to transport the wounded, massive sand dunes, and widely varying march distances. Proper logistics were hugely important to Alexander the Great's successes and failures, and history shows that more often than not Alexander's logisticians, including Hephaestion, often calculated appropriately.[3] While it is tempting to explore the depths of how important mobility logistics were to ancient armies – research on the extensive Roman road system is certainly plentiful – we can make some clear connections between what logistics meant for the Greeks and what it means for current scholars.

These days, logistics is still a field of study that makes use of a wide range of knowledge from other disciplines, including statistics, mathematics, computer science, transportation, and information technology (IT), among others. The field focuses on the arrangement and delivery of people, goods, and services around the world, with a recent emphasis on the role of software, databases, and algorithms in this process. Students in logistics are trained in management, customer service, supply chain concepts, product life-cycles, network design decisions, inventory management, Total Logistics Cost analysis (TLC), Third-Party Logistics (3PL), Intelligent Transportation Systems (ITS), cost models, distribution strategies, and vehicle scheduling. These strategies and systems structure a range of human activity, and most of the myriad decisions that make up logistics calculations are hidden from everyday view, just as most loading docks are only found by accident when we take a wrong turn out of a parking lot. A very common example of the material side of logistics is one we rarely notice, because it concerns the shipping and delivery of goods and can often be found on those same oft-overlooked loading docks.

Though we seldom ever see them, over two billion pallets are in use in the US every day. Use is so widespread, that the production of wooden pallets comprises more than 46% of total US lumber production (Vanderbilt 2012). For logistics workers, one of the largest concerns with pallets lies in "pallet cube optimization," a science that seeks to understand how much of a given material can be feasibly and safely packed onto a pallet. These shipping problems are often the domain of operations research, yet many of the considerations that go into pallet design have to do with specific circulation and delivery concerns: whether the pallet will be re-used, the conditions of storage and transportation, what the pallet will carry, and where the pallet will go. Accordingly, there still exist six different International Standards Organization (ISO) standards for pallet size (Standard 6780) and three main kinds of pallet (stronger, block, and perimeter base). Often in international shipping, pallet sizes will not match up with shipping containers, forcing workers to un-pack and re-pack goods from one pallet to another as they switch containers or methods of transportation (Vanderbilt). My point is not that these logistics are complex, dependent upon complex math, or purely "logical," but that even among the world of pallets, there exist widely varying strategies for material delivery. Those of us who have multiple options for our trip to work in the morning make choices that derive from similar forms of calculation.

The term "logistics" might seem to be an odd choice as a rhetorical term – after all, thc optimization of pallet space and streamlining of delivery solutions does not immediately suggest a sense of rhetorical delivery. However, logistics is a more appropriate choice for understanding persuasive transportation choices than, for example, the terms "network" or "infrastructure."[4] The term "network" is an appropriate description of the de-centralized distribution of potential actors that can be enrolled in various combinations for various purposes. It is out of these networks that specific strategies for movement are developed. However, "network" could easily be confused with the theorized distribution of people and things in constellations of influential actors as described in actor–network theory. The term "infrastructure" refers to the many nonhumans that afford potential paths through the world (here I am thinking of gas stations, roadways, and streetlights). Yet, "infrastructure" also suggests a static structure underlying a dynamic world of action. Instead of a concept wedded to fixed structures, I prefer terminology that suggests dynamic, interactive, and persuasive relationships, as we see below.[5]

Defining logistics

In order to identify how logistics are persuasive, we first need to look beyond modern-day understandings of the concept, away from ancient military logistics as described above, and pursue the meaning of the Greek *logistikos* (λογιστικός). Our understanding of the term is based on a limited, though surprisingly coherent range of definitions from Plato, Aristotle, Diogenes Laertius, Xenophon, Plutarch, and the geographer Strabo. For example, there are 21 instances of *logistikos* in Plato's *Republic* and eight in Diogenes Laertius' *Lives of Eminent Philosophers*. Most of the uses of *logistikos* in these works correspond to the definition given in Liddell and Scott's *Greek-English Lexicon*, that is, one who is "skilled or practised in calculating" or "practical arithmetic, the art of arithmetic" (1996, 1056). In this definition, *logistikos* refers to either the skill of calculation or a person who has been trained in the calculative arts. Plato will sometimes refer to this person as a "reckoner," but will also use the term to refer to those imbued with logic or reason. We might be surprised to learn, then, that Plato offers a definition of *logistikos* in *Gorgias*, hidden in plain sight in one of the most studied Socratic dialogues.

In the dialogue of the same name, Socrates asks Gorgias what his art is, and Gorgias replies that his art is rhetoric and that he is able to make others into rhetoricians, because rhetoric is teachable. Gorgias admits that rhetoric does not treat all kinds of discourse, yet explains that the art of rhetoric allows men to understand about which they speak. Gorgias explains that rhetoric has discourse as its main content, while the other arts have – as their content – the production of something through external action or something worked by hand. Socrates then elaborates that the other arts are "for the most part of making things and that call for little speech," because that kind of work occurs in silence (450c8–9). He argues that in painting or statuary, for example, artists may have no reason to speak, so we might call these arts "silent" arts. Socrates suggests that, unlike these silent arts, rhetoric is similar to arithmetic, calculation, geometry, or draughts (a game like checkers) in that these arts all require a language of some kind. He then claims that these symbolic arts "perform their whole task by means of speeches and call for practically no physical work besides, or very little of it" in order to be completed (450d4–6). Socrates' character defines the logistic art as "one of those arts that exercise their influence entirely by speech" which "differs from arithmetic insofar as computation examines the quantity of odd and even, both in relation to themselves and in relation to each other" (451b5–c4). Gorgias agrees to all of this, and most importantly, also agrees with Socrates when he argues that these other arts – arithmetic, calculation, geometry and the like – are not rhetoric.

This specific section of dialogue in *Gorgias* is important, not only to rhetoricians concerned with where Socrates and Plato differed from the Sophists, but for understanding logistics. Perhaps first and foremost, this dialogue creates a break between the supposed truth-value (and therefore the supposed good) of arts like rhetoric, which Socrates and Plato believed to be a suspect system of knowledge, and the truth-value of other arts like arithmetic, logistics/calculation, and geometry, whose validity was understood as being measurable. Such a conclusion underscores Socrates' point that rhetoric has no real content to be taught or learned, while the other language arts have as their content observable things in the world that correspond to their symbolic language systems. For example, we assume arithmetic helps us add numbers whose representations exist in real life, geometry helps us measure objects in the world, and dialectic and philosophy help us discover and then represent essential and unassailable concepts. As a result, Plato, and presumably Socrates, established what was and what was not valuable among the language arts. This division downplays the usefulness of the rhetorical arts and removes entire symbolic arts from rhetorical consideration (arithmetic, logistics/calculation, and geometry). However, when we examine the Greek term *logistikos*, we understand how the concept is much more than Plato's limited notion of a calculative art expressed by a non-rhetorical symbolic system.

Aristotle does not employ the term *logistikos* often – only a handful of times throughout his work. In book one of *Rhetoric* (chap 10, sec 7), book eight of

Eudemian Ethics (1246b), and in several places in *Virtues and Vices* (1249a, 1249b, 1250a), he uses *logistikos* to mean "rational," often in discussions of reason and "goodness." But, even more than the "rational," Aristotle also offers us the definition of *logistikos* as a *human logistical faculty*. In *Nichomachean Ethics*, he suggests that the soul has two parts, the rational *logos* (λόγος) and the irrational *alogos* (ἄλογος). Of the rational part, we have two faculties, scientific *epistemikos* (επιστεμικος) and calculative *logistikos*. Explaining that "since calculation is the same as deliberation, and deliberation is never exercised about things that are invariable [. . .] the Calculative Faculty is a separate part of the rational half of the soul" (1139a). In this sense, *logistikos* is something we have access to through a faculty[6] that emerges along with the variable situations, contexts, and worlds in which it is a part. Because *logistikos*, as a faculty for situational variability, can only be developed in a material context, it is deeply connected to the humans and nonhumans that comprise our environments.

If we make the connection between the faculty for logistics that Aristotle suggests here, and recent conceptions of delivery in rhetorical theory, we recognize that both identify delivery as a situated ability. Ridolfo and DeVoss' conception of rhetorical velocity, for example, understands delivery as it emerges from a rhetorical strategy, where knowledge of how communication travels across physical and virtual networks helps us determine what kind of delivery strategies to use. Logistics, instead of concerning the transmission, reception, and circulation of a text, concerns movement, interaction, and circulation through material space. Insofar as Aristotle explained that logistics are a calculative faculty, we might call humans *calculating beings*. To say that we are "calculating" means that we conspire, within our unique constellations, to coordinate our world. Here, calculation is more than rules for travel, but our coordination with the myriad interfaces with which we must interact, as we move about. We see the end result of this coordination as a symbolic art and we can call it an art of *logistics*. I should pause for clarity here: logistics is an ability and strategy for coordinating with humans and nonhumans through material movement, and connecting interfaces through movement. While logistics designers can organize an array of potential encounters, we cannot really call something a "logistic" unless it has been brought into life by users. The following example should express how a *human faculty for logistics* differs from a logistic itself.

Founded in Cambridge, Massachusetts, ZipCar is a car-sharing service – different from a car-rental service in one important way: instead of renting cars for solid 24-hour timeframes, car-sharing services rent cars for only a few hours (or as little as 30 minutes). ZipCar requires a free membership that comes with an access card that will unlock a vehicle at a specific ZipCar lot. Each ZipCar vehicle is equipped with radio frequency identification (RFID) transponders that coordinates with the member's card, unlocks the vehicle, records mileage and usage, and makes the vehicle inoperable if stolen. Reservations for individual cars can be made over a Web browser or through a smartphone app, and when a user takes out a vehicle, that person must return the vehicle to the same location. Car sharing services like ZipCar provide increased levels of flexibility at the cost of one-way driving and long trips, so the kinds of logistics they enable are likely to be urban or suburban excursions, possibly for people who do not own a car or who need a vehicle for trips when public transportation would be less feasible. These services are representative of logistical thinking in that they are a calculation of how to get from point A to point B, often involving personal evaluations of cost and convenience. Yet, aside from the calculations that we all do to determine what kind of transportation is most efficient, safe, and cost-effective, a transportation decision must also evaluate the possible variables involved in a specific journey. That is, the heart of what makes car-sharing persuasive for users is not the technology of the automobile, but a transportation option that presents a greater amount of convenience for a smaller amount of complexity among those variables. For someone going to an out-of-the-way store, a car sharing service could be more compelling than, say, a bus or subway. Thus, a car-sharing service enables a calculation of how to move through the world in a way that makes the variables of transportation more agreeable to the users with specific needs and values. Because it offers up a new way of making these connections, it engages our faculty for logistical thinking.

When people envision logistics, they are likely to think of the myriad technologies, people, protocols, and algorithms that go into making a series of connections, like those described in ZipCar, function. That is certainly understandable, because so many of these aspects of a logistic need to work in coordination for people to use the service. That said, a rhetorical conception of logistics understands it as a whole, as a persuasive series of connections that allow for compelling use. For people, logistical thinking manifests as an ability to recognize trajectories through constellations of interfaces, coordinate with them, and then perform a kind of delivery in the world. While it may be difficult to see a specific logistic unless all of these connections and interfaces are charted out for us, or unless we engage in them, most of us would have little difficulty understanding that in order to arrive at a destination, we must make choices about how we would like to be delivered. In ideal situations, we ask ourselves whether we should walk, ride a bike, drive a car, take the subway, or get a taxi. Each logistic brings new variables into play along the trajectory of movement, some of which are more compelling to engage in than others. If we ride a bike, the weather and availability of safe routes play a larger role; if we drive a car, we need to negotiate gas stations, parking lots, and traffic; if we choose a subway, we need to consider ticket prices, the subway map, and train changes. Thus, for humans, logistics are strategies for physical delivery.

At the heart of a logistic is not a technology, a specific kind of space, or even a spatial awareness, but the way that the many different actants, including humans, play roles in a given trajectory and are gathered together. Essentially, a logistic *gathers*. When we participate in a logistic, we particulate in that gathering, because our engagement with many other actants performs the delivery and brings a logistic into existence. For example, the last decade has brought about new manifestations of transportation, including what are known as "transportation network companies" or "charter party carriers" (Geron 2012). These include ventures such as Uber and Lyft and they differ from traditional transportation companies in that they own no large stock of transportation equipment. Instead, they make use of existing application programming interfaces (APIs) within a smartphone application that allows users to hire drivers who use their personal vehicles as part-time taxi services. In busy cities, users log in to the app, which identifies where they are, and choose a driver connected to the service. The map function on the app allows users to see where the driver is and, after they are picked up and delivered, the service automatically transfers money to the driver. When a person chooses Lyft, for example, she brings together the elements set up so as to make a trajectory possible, and causes that logistic to gather through the act of movement. While the *possible* connections need to be in place for potential users, the company that develops a transportation logistics project does not actually see their logistic function unless users perform that logistic and actively *make those connections*. For an app-based taxi service, the pieces are put in place for users who value convenience, speed, and minimal interaction. While these new transportation networks make use of many of the same actants as other logistics, they gather those actants differently. Time will tell whether these services will survive, and whether large numbers of people will sustain the logistic by continuing to coordinate the many interfaces and actants involved.

When we find a logistic to be compelling, such as that of a car-sharing service, then we identify that trajectory as one we would like to continue making. When we find trajectories that serve our needs, it can be surprisingly difficult to alter our routes, as they can become sedimented into habit. A range of critics have commented on the kinds of paths we generate as we move through urban space, though Kevin Lynch, in *The Image of the City* (1960), suggested that these paths are "channels along which the observer customarily, occasionally, or potentially moves. They may be streets, walkways, transit lines, canals, or railroads" (47). Our choice of a commute logistic represents a strategy that we practice again and again. Often, we consider this trajectory as one of the least important and least thought-about moments of our day. In much the same way that metaphors can be deeply compelling when we stop paying attention to them, logistics become deeply impactful when they become habit. John Urry describes this kind of habit as a path dependence that is linked to powerful mobility systems like automobility.[7] He argues that we become "locked in" to different mobility regimes through logistics that:

> linger over time. There can be a powerful spatial fixity of such systems. New systems have to find their place physically, socially, economically and discursively within a fitness landscape in which there are already physical structures, social practices and economic entities that overcome distance and structure mobility in sedimented or locked in forms.
>
> (2007, 53)

In most cases, we are not "locked in" to particular paths, though there are many people in the world who have no choice but to engage in one logistic, because they lack alternative pathways. Some are relegated to taking a dangerous ferry when they would prefer to take a train, or walking when they would prefer to ride a bus. Often, we become aware of logistics when we realize we no longer want to take part in them, or when we find the interfaces awkward to use or gradually less compelling. For example, we may question a logistic when fears over range anxiety keep us from buying an electric car – though this scenario has troubled the electric car logistic long before the recent Tesla controversy.

Early electric vehicles

When one logistic fails to energize users, we can say that it is less persuasive than other competing logistics. In most cases, a lack of accessible, user-focused interfaces can cause it to be less compelling. In the early twentieth century, the logistic of electric vehicles became less persuasive for users because several different actants were not available when needed, while others were difficult to use. These first days of the automobile are best characterized as an overabundance of invention, a scarcity of coordination, and a number of experimental logistics. That is, teams of designers and engineers were developing unique solutions to the desire for motorized mobility, though few of these inventions were coordinated with each other. The general mood of the time was one of searching in the dark for functional and desirable solutions.[8] Gijs Mom, in The *Electric Vehicle: Technology and Expectations in the Automobile Age* (2004), explains the context for this design work:

> The early stage of the automobile was dominated by a hodgpodge of car builders, searching and groping in all imaginable directions, just as had been the case with the streetcar. For the Paris-Rouen race (1894) [for example], the field of expectations was stretched to its limits, and even beyond, because the 102 initially registered entrants were announced to be not only powered by electricity, steam, and gasoline, but also by compressed air, hydraulic engines, "pedal engine," and even compressed water, multiple levers, "combined liquids," and gravity.
>
> (19)

Though "gravity" represented perhaps a less-than-stellar understanding of the general limitations on vehicle design, each of these propulsion schemes represents a unique attempt to understand the technological mobility problem at hand.

Of these numerous technological processes, three main approaches to automobile development and coordination emerged, and each centered around a different method of propulsion: fossil fuel internal combustion, external combustion steam drive, and battery-electric power. Further, each of these vehicle designs relied in part upon vastly different constellations of actants – not only for their creation and composition, but also for their sustained use. Internal combustion vehicles coordinated gasoline stations, hand cranks, hydrometers, and gearboxes; steam focused on clean water, burnable fuel, and flash steam boilers; and electric cars energized central charging stations (which provided electricity to municipalities), exchangeable storage batteries, and charging plugs. All motorized vehicles worked in coordination with drivers, pedestrians, roads surfaces, existing horse, train, and bicycle transport, and so on. While popular histories of the automobile suggest that internal combustion became the dominant propulsion system because it provided drivers with a larger range of operation and more speed, each mobility logistic could lay claim to local success. For example, in the 1900 census, US citizens registered 1,681 steam, 1,575 electric, and 936 gasoline vehicles. While gasoline-powered autonomous transport would shortly emerge as a dominant system, individual geographic locations showed preference for different kinds of vehicles. In 1900, 65% of registered vehicles in Chicago were electric, in 1902, 88% of Portland, Maine's vehicles were steam-powered, and in 1904, 61% of Cleveland's automobile-owning residents favored gasoline cars (Kirsch 2000, 172). Different urban locales enabled different logistics to flourish, even if only for a short time.

As time passed, users encountered a range of logistics, and electric vehicles did not fare particularly well in comparison. In *The Electric Vehicle and the Burden of History* (2000), David Kirsch argues that for residents of New York City, potential electric vehicle owners were confronted with some significant logistics choices. He writes, "[i]f they desired competitive service for a limited range, urban-based vehicle, the local charging infrastructure was too spread out; yet if they wanted minimal service but extended coverage for possible electric touring, the charging stations were not widespread enough, nor did stations support routes between important destinations" (179). Users were not engaging in only one interface, but chains of coordinated interfaces, each with its own benefits and limitations. At the time, electric vehicles did not have the appropriate amount of charging stations to satisfy users' logistics thinking. Though many automobile historians will point to the storage battery's limited range as the determining drawback to early electric vehicles, in reality, "we use energy in the system context," not as one technology in isolation (Kirsch, 24). Instead of blaming one actant (the battery), we can instead see that the logistics of the early electric vehicle resulted in less-than convincing option, at least when compared to other forms of transportation.

In part, electric vehicles failed to become the dominant autonomous transportation technology, or even a viable option for public purchase after World War 1, outside of specific commercial applications, because of a lack of persuasive coordination between humans and nonhumans in the logistic. We can identify a major source of this frustration, too. While steam and internal combustion vehicles could both boast generally homogenous sources for their propulsion (gasoline and clean water/burnable fuel, respectively), forms of electricity for electric vehicles differed greatly. Though such a claim sounds bizarre, electricity came in different interfaces. The lead-acid and Edison iron-nickel batteries used by vehicle manufacturers operated at different voltages, which, when combined with similar lack of standardization in cell number, necessitated the use of in-line rheostats at the central charging station to bring down voltages before electricity entered the vehicle. Once this was accomplished, the owner had to determine which of the eight different charging plugs in use fit the car in question. The owner could wedge pieces of copper into the plug in order to make it fit, but may have been hesitant to do so, because nearly all plug designs had been condemned by municipality fire underwriters (Kirsch, 105). Of course, finding a central station that was willing to charge an electric vehicle was another concern. Though some "electric hydrants" existed, electric vehicle drivers were almost entirely dependent upon their local power plants for electricity, and few of these were interested in supporting electric vehicle charging.[9] Add in a lack of standardization in battery storage facilities, and it becomes clear that early electric vehicle logistics failed to help users gather interfaces or develop a sustainable faculty for electric vehicle logistics.

Mental maps

One of the reasons that electric vehicles failed to remain a popular transportation option in the US for the twentieth century, was that their interfaces were complicated, unsafe, or just simply unavailable. Their logistics did not take hold in the minds and habits of users, in part because not enough users found them compelling enough to engage in on a habitual basis. Human users of automotive transportation are not just the final element that fits into the system at the end, but key components of the logistic itself. In automobile logistics, users are often the most dominant actor – the one that gathers other actants and performs a delivery in the world. Many still feel that electric vehicles lack the potential to become ubiquitous, in much the same way that ICE automobility has. One way we are able to recognize when a logistic becomes habitual is through the development of a more observable version of Aristotle's conception of logistics as a human "faculty."

We are able to tell that a logistic has been successful, on some level, through the development of mental maps. A mental map, or what has been known as a cognitive map, imaginary map (Trowbridge 1913), topological schema (Piaget and Inhelder 1967), cognitive collage (Tversky 1993), or cognitive atlas (Hirtle 1998), is a representation of spatial memory about the ways in which humans make sense of their physical environment. It is thought that mental maps are generally metaphorical, and that they reside in our subconscious mental functioning. The fields of environmental psychology, cognitive science, and geography often refer to the concept of mental maps, usually referring to Edward Tolman's "Cognitive Maps in Mice and Men" (1948) as the first explicit research on the concept. One fundamental inquiry has been to question what form mental maps take and how individuals employ maps for making movement-related decisions. We can, if we like, even identify where in the human brain that mental maps occur. Along with her colleagues, Eleanor Maguire et al. (2000) used MRIs to investigate the brains of London taxi drivers, looking for significant differences between that population and control populations. They found that these taxi drivers had larger anterior hippocampi – and that the size correlated with how long they had been working as taxi drivers. Further research (Maguire et al. 2006) showed that London taxi drivers also have less ability to acquire new visual–spatial habits, perhaps because they developed such nuanced mental maps.

Regardless which region of brain matter is responsible for housing our mental maps, they function as a kind of sociocognitive manifestation of the human faculty for logistics. Often depicted as a *rational ability* (or Aristotle's "rational faculty"), they are highly variable figurations that develop *along with* our active engagements in the world. Mental maps are also proof that a logistic has been compelling enough to encourage habitual movement. We can identify specific interfaces within a logistic that encourage or discourage repeated interactions and therefore movements, too. Kevin Lynch (1960) explains how urban dwellers create and navigate by mental maps and how a city can be "legible" to the people that live in it. Lynch suggests that people make use of many kinds of visual cues, including "color, shape, motion, or polarization of light, as well as other senses such as smell, sound, touch, kinesthesia, sense of gravity, or perhaps electric or magnetic fields" (3). While his work is bit literal in its understanding of how urban designs create affordances, and how those affordances are read by people, Lynch's explanation of the legible city underscores how a mental map is a production of human and nonhuman interaction. For human users of transportation, a city, or any space, for that matter, is made "legible" through the performance of logistics and the eventual development of mental maps. A mental map is thus the manifestation of an engagement in logistics, which is necessarily comprised of a wide array of humans and nonhumans, gathered together to facilitate particular trajectories of movement.

Logistics and technical communication

For technical communicators working on transportation projects, the complexity of planning a logistic, with the expectation that people will engage and perform that logistic, can be daunting. Yet, the larger takeaways from this chapter should reinforce the values many technical communicators already appreciate and the work many already perform. While a range of transportation engineers and designers can establish technologies that will eventually play a role in a logistic, those technologies are only gathered together when in use, and human users hold much of the agency in that gathering. As symbolic-material workers, technical communicators can help logistics designers understand the relationship between users and the materials used, assist in giving voice to those users, and lead efforts to bring those voices to the rest of the team. This role would go beyond developing signage and instructions, for example, but consider the position of users as active agents in the kind of gathering that happens when transportation takes place. Certainly, the methods that we currently employ to understand users – participant interviews, usability testing, ethnographies, etc. – when employed in transportation designs, will help investigate the likelihood of logistical acceptance. We might even argue that technical communicators are in an ideal position to facilitate in this kind of work, as symbolic-material workers that are sensitive to the relationships between users and nonhumans users encounter. As such, they are able to facilitate logistics designs and assist populations less addressed by technicians and designers.

One way in which we can pursue logistics design work is through Madeline Akrich's (1992) research on the relationships between humans and technical objects. She addresses how technical objects bring together "actants of all types, and sizes, whether human or nonhuman" (206). Akrich summarizes how technologists (her broad term for those that envision, develop, or code various technologies) make hypotheses about the world into which their technologies will function, and inscribe this vision in the technical objects they design. She calls this process "inscription" and the result of this work a "script." "Thus, like a film script," Akrich claims, "technical objects define a framework of action together with the actors and the space in which they are supposed to act" (208). The method she suggests to understand this inscription process is to "de-inscript" a technical object, which can be done by going back-and-forth between the designers' *projected* users and the *real* users. Technical communicators working on transportation projects would be working in this inscription process, as it happens, in order to generate *kinesis* forms of movement. Instead of performing a de-scripting analysis on a transportation project, to see where the inscription occurred, they work along with subject matter experts to build the script itself. The value of bringing in users to test and re-test the transportation design lies in giving voice to the users, understanding whether or not they will engage with the logistic, and determining whether the logistic will be successful. While "scripts" are built from data about populations and their habits, the stories hidden in that data represent the complex ways in which people are persuaded to engage in particular paths.

As symbolic-material workers, technical communicators should have awareness of these scripts as they are being formulated and should advocate for users in at least two different ways. First, they can understand and negotiate the nature of these scripts. For example, in a car-sharing service, if people experience nearby parking lots, easy-to-negotiate GUI interfaces, a simple reservation process, a high sense of autonomy, and reasonable prices, the logistic will likely satisfy the users. However, while some users will want hybrid or all-electric vehicles for their short journeys, others may be confused about newer driving technologies or have "range anxiety." Technical communicators would want to help communicate and possibly alleviate these concerns. For all-electric vehicles, users may need to be reassured that their vehicle has enough charge and service representatives may need to check to see whether vehicles have been plugged in after use. By identifying users' concerns, needs, actions, and reactions, technical communicators are able to better advocate for how the logistic will be engaged and aim to make it more persuasive and sustainable for those users involved. A second focus would be to advocate for multiple scripts whenever possible, to avoid the kinds of mobility poverties that currently exist. Users should not be confined to a singular logistic that is overly expensive, inconvenient, or environmentally harmful. When people are given options for how to achieve their daily routines, and when those logistics are made more fitting to their needs, more people are likely to take advantage of them. While technical communicators may not be able to overcome geographic or economic barriers to multiple logistics options, advocacy can occur even without guaranteed success. We can, however, be more vigilant about where mobility logistics can be persuasive, sustainable, and plural.

These positions for technical communication work are going to become only more important in the coming years, as the pressure on logistics will only increase. Though there is no true measure of an increase in the value of logistics planning, some have put forth educated assertions about its increasing importance. Nigel Thrift (2004) has suggested that logistics, which he defines as "a set of knowledges synonymous with movement, effectively the science of moving objects in an optimal fashion," is quickly becoming "the central discipline of the contemporary world" (589). In order to understand the logics behind this rise in "calculative thought," Thrift suggests that we identify multiple different ways that these calculations are performed by different cultures and people with different experiences. We might see a similar value in technical communication work in transportation projects, with the goal of recognizing how different populations of users engage in calculative thinking. The many mental maps that users generate as they engage in different logistics help us understand where users differ and how they perceive new or emerging transportation solutions.

This chapter has attempted to break down the conception of logistics and shift its definition from simple calculation to something that is a performed, engaged habitual movement. I explained the connection between logistics and persuasion, where both humans and nonhumans function together in order to create the possibilities for mobility. Here, logistics are related to the traditional rhetorical canon of delivery, though we might more accurately consider a faculty for logistics as a kind of strategic thinking about delivery – how and where we are best delivered, according to our own personal needs, desires, and situations. Some of this logistical thinking is logical, though much of it is not. Yet, when we engage in logistical thinking, or logistics themselves, we instantiate a gathering of the humans and nonhumans that have been set up to provide the opportunity for transportation. Whether we continuously engage in those trajectories of movement determines the success or failure of a particular logistic. When we consider how certain logistics, such as that of ICE automobility, have become a detriment to human safety and the environment, we see a need for technical communicators to facilitate more environmentally sound, safe, and pleasing logistics options. In the following chapter, I move from conceptions of logistics to the embodied material movement of navigation.

Notes

1 Herodotus narrates this battle in book seven of the "Xerxes" portion of *The Histories* (2008).
2 Records of logistics exist throughout *Anabasis of Alexander*, though Arrian's account of Alexander the Great's army is primarily within book six (2013).
3 Though Alexander is Aristotle's most famous student, Hephaestion was likely his student as well.
4 Mobilities scholar John Urry suggests "system," because "systems make possible movement: they provide 'spaces of anticipation' that the journey can be made, that the message will get through, that the parcel will arrive" (2007, 13). Yet these spaces of anticipation are products of both humans and nonhumans and are only made salient through the performance of both through movement. To that end, I believe "system,"

suggests a thoroughly planned and constructed order, though does not suggest the inherent variability within those sites of order.

5 A much more nuanced and complex understanding of infrastructure has been elaborated elsewhere, perhaps most effectively in Steven Graham and Simon Marvin's *Splintering Urbanism: Networked Infrastructures, Technological Mobilities and the Urban Condition* (2001).

6 I am aware here, too, of the temptation to dismiss Aristotle's conception of calculation as a human faculty as nothing more than a premonition of what would later be called "faculty psychology." Despite the rhetorician George Campbell's work connecting faculty psychology and empirical research with rhetoric, Aristotle's suggestion is not something so reductive as appeals to innate human desires such as "understanding, imagination, passion, and will," as Campbell suggested (Enos 1996, 45).

7 One way to identify when we have become path dependent is to think of how we explain our logistics choices. If we begin dropping nouns out of our logistical thinking, it is likely we are locked into a habit. For example, if I say, "I'm going to the store," but neglect to mention how I am getting there, this path has likely become habitual.

8 I am thinking here of Janet Atwill's description of *techne* as able to "create paths in uncharted territories – to help one find one's way in the dark" (1998, 68).

9 "Despite the larger trend toward load building and marketing new applications for electric power, the electric vehicle market did not generate positive results for electric vehicle station balance sheets until at least 1907. Previously, the automobile charging market has been disappointing" (Kirsch 91). Some battery standardization was introduced in 1912.

6 Navigation

Cruise control navigation

The previous three chapters focused on transportation design, interface development, and persuasive logistics, respectively. In order to build upon that framework, this chapter considers the activity of navigating the world with mobility technologies, something that was referenced in Chapter 5, when I suggested that logistics are only realizable when performed. Now I shift focus to the rhetorical aspects of that complex, human–nonhuman hybrid performance. The example that follows, that of cruise control, illustrates the importance of smart technologies and how those technologies share agency for navigation and use. Cruise control demonstrates several different, yet related concerns with automobility. First, cruise control lets us see that, along with the design of mobility technologies, the act of *using* a mobility technology is also an art. Second, adaptive cruise control's features showcase the unease we tend to feel while navigating with mobility technologies that constantly assert their agency. They complicate what we might understand as a solely human trait – bodily intelligence – as their extended sense of the environment becomes part of our own abilities and limitations. Third, while semi-autonomous technologies help sell vehicles, motor vehicle companies have had difficulty positioning these features in relation to existing driver assumptions. Adaptive cruise control, in particular, carries with it unique advantages and disadvantages.

Cruise control was initially developed outside the realm of automotive transportation. In the late eighteenth century, as train travel became more widespread, engineers had been looking to control speed with greater consistency. While several concepts were developed, one efficient solution controlled engine power through a type of centrifugal governor. James Watt (of Watt steam engine fame) and Matthew Boulton developed a device that involved a spinning "flyball" powered by the steam engine itself – as the flyball spun, it could be adjusted to reduce the aperture of a throttle valve, thereby restricting power to the engine and thus the speed of the train. This was the first real attempt at "cruise control." Similar devices were later employed in automobiles during the early twentieth century, though few roads permitted long-distance automobile cruising at the time. Modern automotive cruise control is generally attributed to Ralph Teetor, an engineer, blinded at a young age, who was supposedly annoyed by how those driving him would speed up and slow down on highways. Initially called "auto-pilot," this system used a vacuum in the car's manifold in order to limit air intake when the vehicle met a predetermined speed (Teetor 1950). First offered as an option on the 1958 Chrysler Imperial, cruise control, and later electronic cruise control (which did away with the vacuum servomechanism), became more popular during the gas shortages of the 1970s, because it helped lower fuel consumption. More significant changes came when automobile makers focused on crash avoidance systems and adaptive and advanced radar-based cruise controls.

Despite their widespread use, cruise control systems have had a controversial history, in part because they represent a shared agency. From the early days of the automobile, people have been concerned about the kinds of emotional, cognitive, and mental sharing that occurs when a driver gets behind the wheel of a car. Specific to long-distance highway driving, the concept of highway hypnosis, or road hypnotism, has been around at least since the first high speed roadways. The term refers to the condition of how drivers can forget time, or focus conscious attention elsewhere, because driving can seem "automatic." The first research on highway hypnosis, a 1929 study by Walter Miles called "Sleeping with the Eyes Open," suggested that sleep might not require closed eyes, especially in cases when "the will to keep one's eyes open is triggered by the importance of a certain task and the personal responsibility involved," such as driving an automobile (Karrer et al. 2005, 455). Recent studies have validated that concerns over cruise control's impact are still warranted. Joint research conducted by France's largest motorway operator and the University of Strasbourg found that when cruise control was used, drivers were less able to merge into traffic, because they took longer to control vehicle speed. Drivers also stayed in passing lanes for longer than necessary, were less likely to move back into the slow lane, and averaged less distance between vehicles while doing so. Further, with cruise control, drivers made fewer corrective actions to their vehicle's trajectory and showed slower reaction times in simulated emergencies. Supporting concerns from Miles' 1929 paper, the 2013 study confirmed that longer travel times also led to an increase in episodes of inattention – roughly 25% more for one hour of driving (*Vinci Autoroutes* 2013).

Of course, even if the mixture of cruise control technology, long-distance highway driving, and humans at the wheel creates dangerous situations, we are not able to point out what is to blame for these dangers. That is, no specific human or nonhuman is at fault, which makes stories like Toyota's recent cruise control controversy so difficult to understand. Fueled by Apple co-founder Steve Wozniak's admission that his Prius' cruise control sent his vehicle surging to 97 miles per hour, those studying the controversy pointed fingers pointed at Toyota's software (Srivastava 2010). In the end, no specific software fault was determined, and Toyota's massive 2009 recall of unsecured floormats (the "pedal entrapment" recall) was unrelated to claims about cruise control acceleration.[1] A more likely explanation is that drivers were not used to operating Toyota's Dynamic Radar Cruise Control (DRCC), which sets vehicle speed based on how closely a Prius follows the car ahead. With DRCC, drivers choose between short, medium, and long "trailing distances" and can increase the set speed by either one or five mile per hour increments using a control stalk. Toyota suspects that drivers accidentally engaged the DRCC, rather than standard cruise control, because it had been the default setting on newer Prius vehicles. It is possible that many drivers did not know this, because the technology is new and few people read the owner's manual (Schweinsberg 2010).

Another model of advanced cruise control is called CACC, or Cooperative Adaptive Cruise Control. CACC-equipped vehicles employ direct radio communication between vehicles and between vehicles and highway infrastructure, thus facilitating a range of driving and vehicle-positioning attributes. The primary advantages to a CACC system involve shorter times for: 1) the perception of an road event, 2) the computation of information to interpret the event, 3) the decision to take action, 4) the selection of an appropriate response, and 5) the initiation of a response (USDoT 2013, 20). The vehicle, working from numerous points of data, can control more agency than humans in specific situations. CACC also allows vehicles to travel at highway speeds with less distance between cars, a situation that can reduce travel time and traffic congestion, and can provide up to a 36% decrease in vehicle emissions with a correspondent 37% increase in fuel savings (6). Yet, for all these potential advances, studies show that when drivers are not engaged in driving as their primary task, their minds wander. Drivers are more likely to engage in "visually demanding secondary tasks," thus reducing their attention on driving (15; 18). Surprisingly, the implications for CACC are similar to the concerns we have had about cruise control for over a century. When we increase the nonhuman agency within the activities related to automobility, with a system like CACC, we find that the balance of agency can be measured through differences in habitual human reaction. Sometimes the overall benefits are positive, though the shared agency can place unusual strains on human users. The act of navigating is a shared agential situation, with both humans and nonhumans responsible for portions of control and intelligence.

In this chapter, I consider how those contexts contribute to an art of navigation that is also an art of use. Here, I mean an "art" in the Aristotelian sense of a *techne*, and in suggesting an "art of use" I mean a *techne* of use. While *techne* is often considered to be more applicable to the making arts, understood as either a step-by-step guide to production or abstract design thinking, it also describes complex use of various technologies. I argue for what we might call a user *techne* and connect such a definition to the myths surrounding the activities of Athena, the Greek goddess of arts, crafts, and skill. Using Detienne and Vernant's (1991) research to explain these myths, and their applicability to the navigational arts, I then connect the rhetorical concepts of *metis*, *tyche*, and *kairos* to research done under the banner of driver–vehicle environment (DVE) studies. Following an explanation of how DVE studies also represent material persuasion, I move to consider the role of agency within increasingly autonomous transportation navigation. While agency is always shared in transportation environments, autonomous technologies, from adaptive cruise control systems to driverless vehicles, bring this shared sense of agency to the foreground.

A *techne* of use

In Chapter 3, I showed how *techne* is most often understood as a making art, how a rhetorical understanding of design relates to this conception of *techne*, and how an Aristotelian definition of *hyle* functions in material design. Yet, this conception of *techne* is only one of several. In Kelly Pender's description, *techne* comes in five flavors: a "how to," a rational ability to attain a useful result, a means of inventing new social possibilities, a means of producing resources, and as a Heideggerrian bringing-forth (16). However, none of these forms of *techne* describes the *techne* which is less demonstrably productive, and which may seem to many to be more akin to "use" than "making." Though Chapter 3 explored both an explicit *techne* 1 and an abstract *techne* 2, a user *techne* does not fit either description. We might even call this a *techne* 3. At first glance, "use" appears much more like an action, or a series of actions, than the complex, contextualized process of production most often identified with *techne*. After all, driving an automobile is a much different kind of *kinesis* movement than designing and producing one. Many rhetoricians might consider any kind of "use," no matter how complicated, to be an action. "Actions" are often connected to practical knowledge (*phronesis*), and thus neither theoretical knowledge (*episteme*), nor productive knowledge (*techne*).

So, where does an articulation of a *techne* more aligned with "use" come from?[2] In part, it emerged in recent technical communication theory through the ascendance of user-centered design practices. User-centeredness, as I briefly explored in Chapter 3, has become a core value for technical communication practice and theory, with Robert Johnson's *User-Centered Technology: A Rhetorical Theory for Computers and other Mundane Artifacts* functioning as a touchstone. Johnson has reflected upon this earlier work, and how the concept of "use" has become seemingly ubiquitous in technical communication. "On one hand," Johnson explains, "the inculcation of user centeredness into such a variety of contexts helped the mushrooming of the concept. This is a good thing" (2010, 337). However, "[o]n the other hand, when a concept becomes ubiquitous, it falls into danger of being used with little historical reflection and concomitant foresight. Such an uncritical use of a concept or practice can render it hollow at best and meaningless at worst" (337). For Johnson, scholars have lost sight of the deep connections between *techne* and use. He identifies the work of philosopher John Wild as significant to this conversation, especially Wild's argument that Plato identified that the "useful end" of any act of making must be part of the process of making. Any art that does not consider "the useful end, on account of which [sic] the art exists" is misguided (Wild 1941, 249). Johnson argues that because the "useful end" is the first of five factors Plato identifies as part of a technical procedure, "the highest of the guiding principles becomes that of use, and use on a social scale" (2010, 347). Basically, Johnson employs Wild's claims in order to show that use is not simply something tacked-on to a technical procedure, but a core aspect of any production or design; Wild claims that the "useful end" of any art should be present in any *techne*. That is, use is essential to any *techne* of making. I believe we can make an even stronger connection between use and *techne* – specifically, that *use is a techne in and of itself.*

We learn more about this side of *techne* in Aristotle's explanation of the scope of *techne* in works other than *Rhetoric* and *Nichomachean Ethics*. In *Physics*, he quite clearly outlines a spectrum of *techne* that includes both making *and* use:

> The arts, therefore, which govern the matter and have knowledge are two, namely the art which uses the product and the art which directs the production of it. That is why the using art also is in a sense directive; but it differs in that it knows the form, whereas the art which is directive as being concerned with production knows the matter. For the helmsman knows and prescribes what sort of form a helm should have, the other from what wood it should be made and by means of what operations. In the products of art, however, we make the material with a view to the function, whereas in the products of nature the matter is there all along.
>
> (194a37–194b8)

Aristotle makes a unique addendum to his classification of *techne*, in that he explains, quite explicitly, that *techne* can also be an *art which uses the product* – one that is directive, perhaps not as formally as *techne*-as-making, and one that is more concerned with form than matter. He claims that the arts that come from contextualized use are as valuable as the arts that concern contextualized making, and one kind of art cannot sustainably exist without the other.

Two scholars who take seriously Aristotle's identification of a user *techne* are Joseph Dunne and Janet Atwill, though they also recognize his examples stretch his definition of the concept. In seeking out the borders of the productive arts, Dunne and Atwill query how Aristotle can identify *techne* as productive knowledge, yet go on to offer examples that do not appear to be productive. In *Back to the Rough Ground* (1993), Dunne explains that this "second kind of *techne* which seems problematic in the light of [Nichomachean Ethics] 6.4's identification of *techne* with productive knowledge [. . .] might be illustrated by such examples as military combat, navigation, or even the important case of medicine" (254). Likewise, in her argument that rhetoric is a form of productive knowledge, Atwill identifies Aristotle's categorization of different forms of *techne* and uses his examples of architecture, navigation, and medicine to do so (2006, 162). However, military combat, navigation, and medicine, while named by Aristotle as *techne*, are certainly difficult to understand as *productive* in the same way that shipbuilding or architecture are *productive*. There is, for example, no easily identifiable product that results from navigation. Given Dunne's and Atwill's references, there appears to be sufficient evidence that Aristotle exemplifies kinds of *techne* as they relate to the use of different material forms in the world – that which he calls "the art which uses."[3]

Dunne argues that (what we can call) a "user *techne*" differs from a "making *techne*" He claims Aristotle implies that "medicine and navigation are *technai* the exercise of which should be deemed to be especially analogous with the living of a virtuous life – so much so, indeed, that *they can scarcely be said to be technai at all*" (258). Yet, Dunne does not consider these to be aligned with *phronesis*, which is roughly translated as "practical knowledge or wisdom." Through an analysis of Aristotle's *Eudemian Ethics*, Dunne finds that:

> Aristotle did envisage a kind of scale in respect of the degree to which different *technai* could encompass the complexity of individual cases within their general rules and prescriptions [. . .] [While] [w]e are left to guess as to how high on this scale productive *technai* (e.g., house-building or saddle-making) might be situated, it is at least clear that medicine and, more especially, navigation are at the *lower* end of it.
>
> (260)

According to Dunne's perspective, one that both Roochnik and Pender support, *techne* is best understood as existing on a spectrum. Rule-oriented activities aligned with the making arts (explicit *techne* 1) sit on one end, broader, more inventive approaches to creation (abstract *techne* 2) are in the middle, and conceptions of *techne* as using art (user *techne* 3) are at the other extreme. Everything on the spectrum is *techne*, just simply different kinds of *techne*. For example, building an automobile and navigating that automobile are both *techne*, though what these actions produce is quite different.

Kelly Pender's articulation of the different varieties of *techne* is especially helpful in understanding how *techne* manifests as production. Her third and fifth definitions of *techne*, as a means of inventing new social possibilities and as a non-instrumental mode of bringing-forth, respectively, align most closely with a *techne* that is less concerned with material production. Pender also finds Janet Atwill to be the staunchest defender of a user *techne*. Atwill identifies *techne* as what occurs when people are asked to use their bodily knowledge and cunning to respond to a particular situation. Far from "ready-made rules for success," this *techne* happens when people engage in complex physical acts and make use of their own bodies, materials, and technologies to perform different actions. Even without explicit rules, as Atwill explains, this *techne* is still teachable, because "it is [still] the character of an art to make its principles explicit and to systematize those principles in order to make them more transferable" (2006, 165). While such a *techne* is certainly less directly productive than the more straightforward making arts, it does produce several things. First, this *techne* produces new social relationships and possibilities because it brings humans together in new and different ways. Second, it can likewise produce new ways of being in the world and new ways of understanding. When we use a particular technology, and build a bodily knowledge through that use, we also develop a way of moving through the world that changes how we see the world. As I mentioned in Chapter 1, any regime of mobility requires an active engagement in those mobility practices. When we take part in them, we not only help sustain them, but also produce a regime of mobility that then shapes other social, economic, and political, and material realities. A user *techne* is a productive art, though instead of manufacturing materials, it occurs with and makes use of material actants.

Navigation as a *techne*

The above scholars suggest that there is a version of user *techne* different from a *techne* of productive making, and that such a *techne* functions by general principles that need to be practiced in material contexts. This user *techne* is more difficulty to identify, in part because there is no specific material outcome to which we can easily point. Specifically, I am thinking of the reference to navigation. Whether people are piloting a ship in ancient Greece or driving a car in modern day America, we could suggest how such activities are productive, though the "production" would not result in a tangible product. These *techne* are better described as user *techne*, as they are more closely related to use than either *techne* 1 or 2. Marcel Detienne and Jean-Pierre Vernant's and Debra Hawhee's investigations into the classical Greek concepts of *metis*, *tyche*, and *kairos* also help us understand how navigation is in fact a *techne*.

Detienne and Vernant center a fair portion of their text, *Cunning Intelligence in Greek Culture and Society* (1991) on the examples of shipbuilding and navigation and their observation that "building and driving are two types of activity in which we are more prone to notice the differences than the resemblances" (234). Building/making and driving/using certainly appear different, even though they may share a common mythological root. Detienne and Vernant explain that in the *Odyssey*, the goddess Athena is responsible for preparing Telemachus' ship as he sets off to look for Odysseus. Not only does Athena build the ship, anchor it at the entrance to the harbor, and choose the pilot, but she also assists with navigating the boat through treacherous waters (215). Athena is a symbol for both effective navigation and the union between land, sky, and sea. She is often depicted assisting with building ships and with their navigation, especially as journeys become complicated by weather, waves, and treacherous shorelines. For Detienne and Vernant, Athena is representative of how the Greeks made little to no distinction "between building and steering, between the cutting of the keel of a ship along the line and steering the course of the ship over the sea" (237). The knowledge, practice, and skill that goes into shipbuilding and the knowledge, practice, and skill that goes into navigation represent different ends of a continuum of *techne*.

For the Greeks, the connection between building and navigation ran even deeper than mythological and symbolic narratives – it is likewise represented in their everyday concepts for direction. What I mean by "direction" is both an instruction to follow in the fabrication of a boat and the navigation that guides a boat through the waves. Detienne and Vernant are helpful in sorting out this language, too, as they explain that the Greek phrase "*epi státhmēn ithúnein*" translates as "to lead straight" and was used by shipbuilders when measuring wood to be cut (237). To cut a piece of wood, for example, a shipbuilder would need to follow a straight line. They explain that "[i]f a carpenter can cut a ship's keel fair and straight with the aid of a line he does this by the good grace of Athena who has endowed him with skill in woodwork" (235). Athena aids in the *techne* of making concerned with producing a ship, however, "*Ithúnein*" is also the word for what navigators attempt when guiding a ship to lead it along a straight heading (237). Detienne and Vernant explain:

> To direct, to correct, to guide a straight course, *ithúnein*, are commonplace expressions in the terminology connected with navigation but by their very banality they stress the importance in the art of the pilot and of the combination of skill in foreseeing where the route lies and the ability to keep one's sights set on one's ultimate destination.
>
> (226)

This connection between ship carpentry and ship navigation is emblematic of how the movements involved in a user *techne* are similar to a making *techne*. Much like how the representation of Athena suggests no difference between building and navigation, the lack of difference "between cutting the keel of a ship along the line and steering the course of the ship over the sea" suggests the same (237).

We are able to map Athena's duties to something even more closely resembling a modern-day *techne* of automobile navigation, because Athena is also the goddess of chariot driving. Detienne and Vernant show that Athena's role extended beyond the specific confines of the sea to the practice of navigation in general, regardless which vehicles were being engaged. Athena gave men the art of chariot building and "[j]ust as she has shown woodworkers how to make ships or ploughs, Athena has taught them the art of building chariots and wagons" (234; 235). Detienne and Vernant conclude, "where the horse-drawn chariot is concerned, Athena's activity [is] not restricted to driving the chariot and horses but extends to the building of the carriage and the putting together of the various pieces of wood from which the chariot as a whole is composed" (234). While Athena's domain encompasses the specific practices of making and use, more importantly, her power is manifest over the spectrum of *techne*. Automobile navigation *techne* is admittedly much different from the specialized engineering and design *techne* involved in developing and manufacturing a car, but both practices represent the full spectrum of *techne* suggested by Pender, Roochnik, and others. Further, both a making *techne* and a user *techne*, regardless of their connection with ship carpentry or navigation, result from, and make use of, the same kind of intelligence – *metis* (μῆτις) (235).

Metis, *tyche*, and *kairos*

Debra Hawhee interrogates the concept of *metis* throughout *Bodily Arts: Rhetoric and Athletics in Ancient Greece* (2005) and defines it as an immersive intelligence and bodily knowledge deeply connected to *techne*. As Hawhee explains, *metis* corresponds to "the ability to cunningly and effectively maneuver a cutting instrument, a ship, a chariot, a body, on the spot, in the heat of the moment" (47). As such, *metis* is a knowledge *applied* within the context of different material forces. The best way to win a chariot race or cross a sea in a storm (difficult conditions, both) is to use crafty knowledge to control horses or align sails with advantageous winds. Most definitions of *metis* reference the concept's association with trickery, and *metis* is often defined as "wisdom, skill, [or] craft," associated with a *cunning* intelligence (Liddell and Scott, 1130). It is often claimed that the wily, intelligent body possesses *metis*, because the best way to negotiate complex or difficult situations is through clever movement and positioning. A range of animals are said to be endowed with *metis*, including the fox and the octopus. The fox is quick and sly and the octopus can outmaneuver foes in order to avoid capture, or slip out of sight, mimicking the color of its surroundings. Yet *metis* is more than about cleverness in avoiding capture. Along with a cunning trickery, *metis* refers to the general mastery of control over a body – a physical, performative knowledge that is employed in a material context.

Because *metis* takes place in bodies, it is also considered the bodily knowledge of *techne*. Aristotle's comments on Empedocles' and Parmenides' treatments of *metis* suggest that it is best understood as "as a kind of intelligence" not "separate from bodily state [*hexis*]" (Hawhee, 57). Detienne and Vernant describe *metis* as existing only at the *moment of use* and argue that "the art of the helmsman can only be exercised within the framework of the uncertainty and instability of the sea. The play of the tiller cannot be dissociated from the movement of the waves" (223). Bodily intelligence is always situated within bodies in material circumstances. Hawhee explains just how rooted *metis* is, suggesting that "*metis*, by its very nature, needs to be deployed – it does not exist on its own, but only in connection with its use" (53). For example, the *metis* of an automobile driver can only be said to exist when the driver is in the car, performing the actions and making use of her bodily intelligence as she maneuvers the vehicle on streets, through traffic, in accordance with various motor vehicle laws. *Metis* is perhaps best defined as an immersive or performed "intelligent ability" of those who perform a range of techniques in response to specific situations (Detienne and Vernant, 11). While the shipbuilder and navigator have different end goals (to complete a ship and navigate the sea, respectively), they both make use of their bodily knowledge to perform their arts.

Since *metis* is emergent in specific, material conditions, it also stands to reason that general rules and strategies for navigation cannot be applied without regard to the particulars of a given situation. Detienne and Vernant show that *metis* accounts for "how the Greeks represented a particular type of intelligence at grips with objects," and focused on interactions between people and material constraints (2). For example, a driver, no matter how well prepared, can never fully know the unique material interactions she will encounter on the road. There will always be variables that cannot be accounted for ahead of time. Janet Atwill cautions that *techne* is in no way a *representational* body of knowledge, but a response to and an intervention in existing conditions. While *techne* includes some forethought (driver education courses, driving tests, practice with parents, etc.) no amount of preparation will trump the importance of the situational confrontation involving another automobile driver, bicyclist, or wayward pedestrian. Hawhee calls this kind of context a "matrix of response" and references Detienne and Vernant's complementary point – that *metis* "is not an explicit set of precepts but rather a tacit style of movement" (Hawhee, 48; 47). While *metis* is kind of bodily intelligence, it is also a mode of response that is then put into movement. Employed in a chariot race, for example, *metis* would be the kind of strategic bodily reaction that changes from moment to moment and helps chariot drivers navigate all manner of material conditions, from tracks and chariot wheels, to horses and drivers' whips.

Because *metis* is enacted in specific conditions, it is also the prime response by which any actants engaged in a user *techne* must confront chance, or *tyche* (τύχη). Defined variously as chance, fortune, providence, or fate, *tyche* is what any *techne* works to control, yet even more than that, *techne needs tyche* (Liddell and Scott, 1,839). In much the same way that *techne includes foresight*, *techne* cannot be said to be fully engaged without a confrontation with *tyche*. Foresight without a confrontation with chance is simply abstract planning, not true foresight, which works against a background of potential instability. A confrontation with chance "brings the indiscernible future within the bounds of possibility," or, to put it another way, a user *techne* takes *tyche* and converts situational conditions and materials into "usefulness" (Detienne and Vernant, 223; Edmunds 1975, 42). *Metis* connects to *tyche* because it "applies to situations which are transient, shifting, disconcerting and ambiguous, situations which do not lend themselves to precise measurement, exact calculation or rigorous logic" (Detienne and Vernant, 3–4). Within a user *techne*, *metis* functions as the active material engagement with *tyche* – the response of the automobile driver, for example, as he confronts the variability of the city and creates a path home.

Finally, a successful user *techne* must not only employ *metis* and work with *tyche*, but also identify and engage *metis*-guided maneuvers at specific and appropriate times, or with what is known as *kairos*. Precise timing is difficult to achieve, but for the ancient Greeks, it was at least commonly understood. Atwill states the basic connection between *techne* and *kairos* plainly, arguing, "if *metis* is the intelligence identified with *techne*, then *kairos* is the time 'associated with techne'" (1998, 57). *Kairos* can be used to reference "limit,"[4] though is most often translated as "exact or critical time, season, [or] opportunity" and generally "marks the quality of time rather than its quantity, which is captured by the other, more familiar Greek word – *chronos*" (Liddell and Scott, 859; Hawhee 2005, 66). *Kairos* has received more attention from rhetoric scholars in recent years (certainly more than *metis* or a user *techne*), especially since James Kinneavy's extensive work on the topic (1986, 1994), as well as investigations from Carolyn Miller (1992), John Poulakos (2002), and others.[5] The ability to seize *kairotic* moments requires a competent *techne*, an employed *metis*, and a confrontation with relevant *tyche*. As a bodily intelligence, *metis* enables the navigator to take (often subconscious) advantage of opportune situations as they arise. And like *techne* and *metis*, *kairos* can only truly be said to exist while it is happening in the immediacy of an event, that is, *kairos* is "immanent, embodied, [and] nonrational" (Hawhee, 71). It is more than a repertoire, but a grasping of opportune moments with an embodied knowledge specifically so that chance may be confronted and the situation may be turned beneficial for the user. As Atwill argues, a *techne* "deployed at the "right time" may do more than redefine the limits of specific situations; it may also create alternative situations" (60). The application of *metis* and *kairos* in a user *techne* shows how something like the art of navigation includes a range of immanent, bodily, time-focused rhetorical traits.

These elements of a user *techne*, and specifically a *techne* of navigation, are applicable in more locations than obscure Greek mythology or debates over terminology in Greek rhetoric. They also appear as relevant considerations in driver–vehicle environment studies. In part, this is because considerations of bodily intelligence, propitious timing, and chance are all relevant aspects to the material context of a driver's immediate environment. Further, they help us comprehend the complexity of material persuasion during a *techne* of navigation.

Driver–vehicle environment studies

Researchers in fields that attempt to predict, analyze, and model driver behavior have described the art of navigation using similar concepts to those I outlined above. These fields are not concerned with persuasive transportation, exactly, though they describe, in different terms, the *techne* of navigation. Researchers have developed a wide range of different methodologies within behavioral and cognitive psychology in order to understand humans in driver–vehicle environments (DVEs). Much of the early behavioral research on DVEs, such as Weir and McRuer (1968), focused on drivers' ability to handle a vehicle and attempted to measure the difference between a driver's stated goals and his or her eventual accomplishments. More recent DVE information-processing and cognitive models offer descriptions and predictions of cognitive activities. Other models have merged what is valuable in older cognitive, behavioral, and psychological models while taking into account concepts such as attention, perception, task-awareness, goal-setting, motivation, safety, risk, and emotion, placing them within movement-oriented hierarchies. For example, Rasmussen (1983) divided tasks into knowledge-based, rule-based, and skill-based levels, Michon (1985) into strategic, tactical, and operational levels and Alexander and Lunenfeld (1986) into navigation, guidance, and control levels. Others like Ranney (1994) have argued that many cognitive information-processing models are a bad fit to real-life driver behavior, because they are incompatible with highly complex scenarios like driving. More recent models have incorporated hierarchical understandings of action into stimulus-response theory. Though DVE studies rely on a variety of these models, and newer adaptations, most are intent upon breaking down what I have been calling a *techne* of navigation into smaller, measurable concepts.

Perhaps most interesting about current DVE models is that they strive toward fine-grained representations of an art of navigation. Specifically, they understand driving as a complex, immersive, and contextual practice where the driver never simply "acts" upon his or her surroundings. Peters and Nilsson argue that cognitive perspectives on DVEs view driving as:

> one of the most complex and safety critical everyday tasks in modern society [. . .] Driving a car is complex in the sense that it requires the driver to employ a wide range of abilities in order to interact with a complex environment and to manage the driving task demands. Driving is dynamic as the demands can

> change back and forth from very low to extremely high, sometimes within fractions of a second.
>
> (2007, 85)

Psychological DVE models also focus on a driver's unique experience level within a traffic system and the creation of "rules and schemas that control their behavior in various (foreseen) situations" (Weir and Chao 2007, 294). Under these theories, drivers are identified as part of their vehicles – at least while they are in motion – reflecting what Haraway has argued about cyborg bodies, Latour claims about hybrid actants, and Tim Dant sees as the driver–car assemblage. Models may assess a driver's cognitive, psychological, or behavioral characteristics, but they also recognize that drivers simply cannot be "extracted" from a vehicle in order to be studied in a representational model. Common terminology refers to this assemblage as the joint driver–vehicle system (JDVS) and sees driver and vehicle as inseparable during movement. *Metis* has likewise been present in a number of models for understanding and predicting driver action in DVEs.

Metis, like *techne*, is never directly referenced in these studies as such. Yet, a focus on what we might recognize as "bodily intelligence" has existed as far back as James J. Gibson and Laurence E. Crooks' influential study, "A Theoretical Field-Analysis of Automobile-Driving" (1938). Gibson and Crooks argued that drivers develop an understood "field of safe travel," for which they define as "at any given moment, [. . .] the field of possible paths which the car may take unimpeded. Phenomenally it is a sort of tongue protruding for-ward along the road. Its boundaries are chiefly determined by objects or features of the terrain with a negative 'valence' in perception" (454). Gibson and Crooks' rudimentary phenomenological model (see Fig. 6.1) theorizes a driver's perception of velocity and the knowledge of what maneuvers a driver–car is capable. This diagram represents the driver–car's perception of how fast it can go, where it can travel, and how other vehicles and traffic conditions will impact braking, accelerating, and steering. Like most early behavioral models in DVE studies, this one attends to the steering and braking maneuvers driver–cars can take to avoid accidents. Gibson and Crooks' articulation of what drivers do also includes nascent conceptions of *both knowledge and ability*, and represents a bodily intelligence that can be employed as situations unfold. Throughout time, DVE studies have added on a wide range of driver tasks to consider, including cognitive tasks not necessarily connected with discrete driving inputs. For example, Hollnagel and Woods' (2005) Extended Control Model describes several concurrent driver processes, each operating on different levels. They identify "targeting" tasks (general goals), "monitoring" tasks (perception of environment), "regulating" tasks (maintenance of safety margins), and "tracking" tasks (small corrections). Each task engages some measure of bodily intelligence, where successful performance requires bodily *metis* developed through experience and training. Whether a driver–car reacts to a gust of wind or subconsciously attends to a car ahead of it on the road, it monitors what movements are possible. DVE studies have generated more detailed understandings of what happens with the driver–car assemblage, but we can see how these detailed models explore bodily knowledge in the art of navigation.

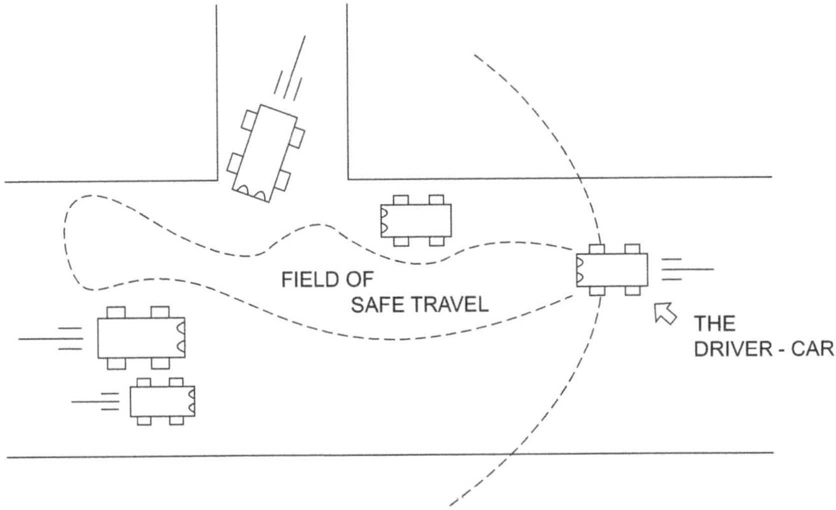

Figure 6.1 A model adapted from Gibson and Crooks' "field of safe travel and the minimum stopping zone of a driver in traffic"

Source: Adapted from Gibson and Crooks 1938, 455

Kairos is perhaps more difficult to locate in DVE studies, though several studies consider "right-timing" when it comes to driver–cars' perceptions of their own and others' movements during navigation. Some scholars render the perception of distance into a measurable number, where these numbers represent the *quality* of time. For example, studies have considered Time-To-Contact (TTC), Time-to-Line Crossing (TLC), Time-to-Intersection (TTI), and Time-to-Stop-Line (TTS). Ostensibly, these measurements are quantitative, but they can also be read as discrete depictions of *kairotic* events. How much time is needed to avoid contact, for example, is a rapidly made bodily estimation based on *metis* and an understanding of the nonhuman factors in an immediate driving context. Again, as far back as in Gibson and Crooks' study, we can recognize a graphical representation of *kairos*. Fig. 6.2 shows what a driver must understand as she is to attempt a passing

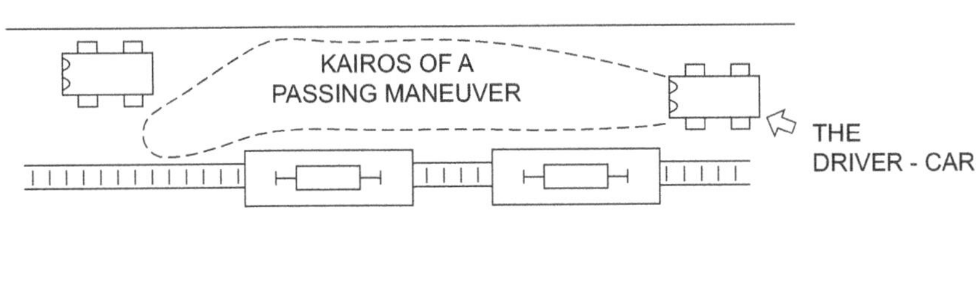

Figure 6.2 A model adapted from Gibson and Crooks' "A driver attempting to pass a street-car before reaching a parked car"

Source: Adapted from Gibson and Crooks 1938, 464

maneuver – the speed of a street-car, the affordances of the road, the distances involved, and the ability of the driver–car assemblage. A sense of bodily knowledge is not enough if this situation is to be safely negotiated. A sense of propitious timing is also required.

The "bit" and agency

As we can see, navigation is a complex series of interactions, none of which would happen, and none of which could be made manifest, if a range of humans and nonhumans failed to function together. The bodily knowledge of *metis*, the chance occurrences of *tyche*, and the propitious timing of *kairos* require each other for navigation to occur – and we can identify these concepts in past and current DVE studies. That navigation is an art that requires the coordination of these concepts brings us to a realization about different user *techne*: they happen because the human and nonhuman actants involved function together in a complex ecology of articulation and influence. Just to start an automobile, we encounter the shape of a key, the tactile response we register from the ignition switch, the specific sounds that cue us that the ignition has functioned, and the various levers and pedals that give additional visual and tactile feedback. Assuming that these actants work together, we then get to leave our parking spot and engage an even wider array of humans and nonhumans. We "feel" the road a certain way through the steering wheel, road signs direct our attention to obstacles and regulations, paths marked on the road suggest trajectories for our driver–car, and we make new meaning with these interfaces to further our movement, all the while altering our trajectory. We make use of the bodily intelligence we have developed with the car we drive, we react to unexpected changes in the road, and we recognize timely opportunities, large and small, as we negotiate traffic, all while pursuing a larger goal related to our specific destination.[6] Because of these decisions, both conscious and unconscious, it may seem as though we are in control over the vehicle we drive, and while this control is rendered to us as a kind of mastery, a huge array of nonhumans are active agents in vehicular navigation.

When engaged in this navigation, we are often oblivious to the vast array of actors that work with us (or sometimes against us) to make movement possible. Many subconscious deployments of *metis* in coordination with *kairotic* opportunities become compelling in part because they are habitual and easy-to-ignore chains of meaning-making and *kinesis* movement. Our consistent use of automobiles to drive to work, for example, is compelling in part because active attention is not required to perform much of the navigation. Of course, different navigation situations ask more (or less) of us as navigators, too. For example, passengers on an airplane are presented with fewer decisions than automobile drivers. Driving provides a sense of autonomy and capability, even though almost all of our vehicular navigations take place within a strongly defined set of rules and boundaries. But, when we assume that all that needs to happen in order for our automobile to function is for us to "take command" of it, as a subject dominating an object, we shut out the more nuanced forms of navigation and the ways in which agency manifests. In the more complex human–nonhuman relations that mask distributed agency – like automobile navigation – the question "who is at the wheel?" is difficult to answer.[7] While lawsuits reach verdicts about who or what is at fault, our everyday automobile journeys are infused with shared, distributed agency.

The assertion that agency is distributed in complex actions, like vehicle navigation, is supported by rhetorical theory and scholars who claim that we do not grab hold of agency so much as we are inhabited by agential forces. As explained in Chapter 2, agency is not something that can be held, but is something that inhabits both humans and nonhumans, depending upon the context. Lundberg and Gunn (2005) espouse this view, and we can make connections between their work and actor–network theory's fundamental assertion that humans and nonhumans share agency in collectives of force. Humans often rise to exert more power and host a greater share of agency, even though the work of humans (and nonhumans) in these collectives is often that of translation, delegation, mediation, and sorting, among other actions (Latour 1993a, 137–138). With automobile navigation, we can easily identify moments when nonhumans are in greater control (for example, when a smart parking system takes over), and times when the human driver is in greater control (when we take action to avoid an accident).

Rather pleasingly, another Greek myth illustrates shared nonhuman agency in the art of navigation. Pindar's *Olympian Odes* describe how after taming and breaking the horse for humans, Athena gave Bellerephon the gift of a horse named Pegasus (Detienne and Vernant 1991, 188). Along with this horse, Athena is credited with giving humans a way to help control horses in general, which were well-known to be temperamental animals. We know this gift as the "bit," a piece of metal, connected to the bridle, that allows a rider to communicate with a horse. Ancient Greeks believed the bit was a living thing that emerged from the *metis* of the blacksmith. Detienne and Vernant explain that the bit is not just a product of metallurgy, but also "a technical object which makes it possible to control a beast of unpredictable reactions" (197). We can see the bit as representative of a moment when the axis of agency in the relationship between humans and horses shifted slightly more toward humans, as a bit is capable of communicating a great deal of information about where a rider wants to go through the application of pressure on the roof of a horse's mouth. The bit is both a symbolic representation of the governance we have in our chosen navigations as well as how we share that governance with other nonhumans. If we want to see what a "bit" would look like with regard to vehicular navigation, as opposed to horse riding, the closest ancient Greek analog might be that of chariot driving.

If we shift from humans and horses to humans, horses, and chariots, the role of the bit is much less precisely defined by a single kind of technology. With a chariot, the agency of the bit is distributed among a wider range of human intelligences, nonhuman material affordances, and animal instincts. The chariot driver, like the automobile driver, may perceive that he has a great deal of control over his movements (as a driver's interactions with the interfaces in the chariot result in new movements), but that is not always the case. In chariot driving, a huge range of actors must function in concert if a vehicle is to be propelled in any direction, a different corporeal situation from, say, walking (the most basic form of human movement, according to Aristotle). Detienne and Vernant explain that the Greeks articulated the difference between horse-riding and chariot-driving as such:

> In chariot-driving it is no longer the bit which confers mastery over the chariot. Here the effect of the bit is much less important in guiding the horse than in horse-riding. Nor yet should we simply identify the reigns (*hēnia*) as the technical object covered by the verb *hēniochein.* Athena's role has wider connotations covering the whole system of behavior which a driver of chariots must put into operation. It involves a quick eye, immediate reactions and paying close attention to unpredictable behavior from the horses, to any unevenness in the track and to all the obstacles which might cause the chariot to veer off course but which a prudent driver, a *hippómētis* one, knows how to turn to his advantage.
>
> (204)

For Greeks, controlling a chariot was much different than controlling a horse. The sensible chariot driver, the *hippómētis,* employs an "art of harnessing a chariot, *hárma zeûxai*" in the "art of driving the chariot team, *hēniochein*" (203). With a chariot, more technology than the simple "bit" is incorporated into movement, and agency is shared among human intelligence, horses, and a range of other material affordances.

Moving from ancient Greek chariots to current and future automobiles, we come to recognize an even greater value in designing with considerations of distributed agency. In Chapter 3, I explained how a *techne* oriented toward making should respect the role of nonhuman actants in the design process, so as to avoid technical designs that fail to fit the contexts in which they are intended. I stressed attentiveness to *hyle*, the Greek term for the general properties of a material in the immediate context of a technical project. I also suggested that electric taxi cabs were successful when designers focused attention on the *hyle* of their projects to fit the geography, population, and other materials available in Manhattan, but largely unsuccessful when they attempted to distribute their vehicles to other cities. In innovative transportation projects, nonhuman actants involved in design matter a great deal, as most play a significant role in the success of that design. Similarly, nonhuman actors play a significant role in a user *techne* and its coordinate relationship with *metis* and *kairos*. The "bit" in user *techne* is not representative of agency conveyed through a single material, but agency distributed through a wide array of humans and nonhumans, including the person in the driver's seat. We can see such distributed agency in recent advanced transportation designs, including perhaps the most talked about new vehicle of the last decade – the Google car.

The Google car

Many examples of nonhuman agency are subtle, though we can point to a particularly distinct example that showcases the nonhuman agents involved in both a *techne* of making and a *techne* of use. In 2002, the US government's Defense Advanced Research Projects Agency (DARPA) announced that it would fund a competition to promote autonomous vehicle research, with the ultimate goal of creating autonomous vehicles for the military. The 2004 DARPA Grand Challenge asked competitors to design, build, and enter a fully autonomous vehicle in a difficult 150-mile race through the desert around Barstow, California, and Primm, Nevada (DARPA n.d.). The 2004 DARPA Grand Challenge was not particularly successful, as no vehicle made it further than 7.4 miles, yet interest was significant enough for the challenge to be held again the following year. Five vehicles successfully completed the 2005 Grand Challenge, with a vehicle fielded by a team from Stanford University taking the win. Stanford's project was headed by Sebastian Thrun, at the time, an Associate Professor and Director of Stanford's Artificial Intelligence Lab. Thrun may be better known from his association with massive open online course (MOOC) developer Udacity and Google's Street View and driverless car program.

The Google driverless car project brings together innovations developed by Thrun and others in the DARPA challenges. As of 2014, the project involved combining laser, radar, GPS, and software modeling with Google's own Maps and Street View databases in order to develop largely autonomous vehicles. Thrun's work with the Google driverless car has attracted a great deal of media attention, but also a number of concerns. The thought of autonomous vehicles fills some with dread, because, as one news article explained, driverless cars would need to be hundreds, if not thousands, of times safer than personal computers, "which crash on occasion and are frequently infected" (Markoff 2010). After all, no one wants their car to come down with a virus and drive off the road while on the way to work. Yet, in both the DARPA race and the development of the Google car, the vehicle itself plays a strong role in the sustainability of a making *techne* and a user *techne*. With modern autonomous vehicles, their active agency is pushed to the foreground.

We can make the argument that both the Stanford DARPA vehicle and the Google driverless car act as agents in innovative *techne* of making and use because they function through a relatively new understanding of software coding, movement, and agency called *probabilistic robotics*. Thrun describes probabilistic robotics as a software-intensive approach to robotics design, where the robots themselves operate in unstructured, contingent environments. As opposed to a robot that spot welds automobiles on an assembly line, an autonomous vehicle must recognize, estimate, and move through a world with an astounding array of variables. There are, after all, more unknowns on a busy city street than a production line. As Thrun et al. (2006) explain, the key to successful navigation with probabilistic robotics "is to represent uncertainty explicitly, using the calculus of probability theory. Put differently, instead of relying on a single 'best guess' as to what might be the case in the world, probabilistic algorithms represent information by probability distributions over a whole space of possible hypotheses" (3). When a vehicle like a Google autonomous car senses the world through multiple systems, each hypothesizing on variables in the world around it, it becomes a data-gathering and data-representing actant, moving through physical locations that humans are unwilling to go or less-willing to actively engage.

A Google driverless car "sees" the world through a number of different components that gather information: a long-range laser system with 64 beams provides three-dimensional depth, a radar system gathers data at a distance, and a camera identifies color, good for understanding road signs, turn signals, and lights (Bilger 2013, 105). All of this data is collected in powerful computers that run a number of algorithms. Some algorithms set basic parameters and hard rules, while others manage the incoming data and make decisions based on the likelihood of certain outcomes. According to Thrun, the vehicles are successful at navigating because of the hypotheses of probabilistic robotics. The central software assumption is that these vehicles anticipate "various contingencies that might arise in uncertain worlds, thereby seamlessly blending information gathering (exploration) with robust performance-oriented control (exploitation)" (2002, 53). A probabilistic robot builds its own knowledge of the world, employing an understanding of its own limitations as it moves. Thrun has no problem suggesting that these kinds of robots *carry a notion of their own doubt about the world with them*:

> *A robot that carries a notion of its own uncertainty and that acts accordingly is superior to one that does not.* In particular, probabilistic approaches are typically more robust in the face of sensor limitations, sensor noise, environment dynamics, and so on. They often scale much better to complex and unstructured environments, where the ability to handle uncertainty is of even greater importance.
>
> (Thrun et al. 2005, 5)

If we follow these assumptions, a vehicle like the Google driverless car could be assumed to have agency in the art of navigation. Google driverless cars must recognize all kinds of road hazards, including other cars, people, cats, tumbleweeds, and empty boxes in orders and positions never before existent. This data becomes aggregated over time, allowing the vehicle to "learn" about its surroundings and how to navigate them. Some might call this process of data gathering just another human technique, though probabilistic robotics' assumption of active engagement suggests that the vehicles themselves have increased agency in navigation. Interestingly, Google vehicles have been particularly effective at this navigation, too, especially if we consider "effective" to mean safe and efficient. In a country where 94% of all vehicular crashes are due to driver error (USDoT, NHTSA 2015), Google self-driving vehicles have been at fault for none of the 14 minor accidents involving other cars (as of the summer, 2015). That statistic is especially impressive, considering the 1.8 million miles that those same self-driving vehicles have covered (Google 2015). The agency that Google's autonomous cars wield, then, seem to be effective in real world scenarios.

Navigation and technical communication

As we move toward greater experimentation with autonomous vehicles, and as we begin to see more integration of autonomous or semi-autonomous features in automobile design, technical communicators working with transportation designers can function in several different capacities. In our ongoing concern for human users, we should understand that few current drivers will be immediately comfortable with relinquishing that sense of agency. The feeling of the "bit," though already distributed throughout an automobile, confers a sense of control to the driver. With many autonomous vehicles, this feeling may be reversed, as the nonhumans appear to passengers to be in control and appear to maintain greater agency as the vehicle navigates the world. While many surveys have sought out public opinion on autonomous vehicles (and the results vary widely from country to country), about 65% of Americans are "very" or "moderately" concerned about riding in a completely self-driving vehicle (Schoettle and Sivak 2014). Similar to how a driving instructor, wary of a student driver, hovers a foot over a second brake pedal, humans are likely going to want to retain some sense of agency in an autonomous vehicle. This concern over independent vehicular agency can take many different forms, but may initially mean that self-driving cars will retain commonplace interfaces such as the steering wheel, accelerator, and brake pedal. As of 2014, states that allow for autonomous vehicles still require that those cars be equipped with such driver-focused equipment. Many companies are aware of this concern, too. Ford's vehicle engineering supervisor for driver assistance technologies has explained that drivers are not opposed to the concept of autonomous vehicles, *per se*, but will need to be helped into gradually understanding autonomous mobility (Fitchard 2012).

As transportation design projects integrate autonomous technologies, technical communicators will be called on to understand the relationships between drivers and nonhumans in the art of navigation. Recognizing the full scope of a *techne* of navigation, usability testing methods may need to broaden, as usability studies will not only need to account for how humans make use of nonhumans, but how an art of navigation is practiced by humans and nonhumans in tandem. This greater integration of autonomous features may require us to adjust usability methods in the driver–vehicle environment, perhaps redefining it as the study of agency in the art of navigation. This shift would involve additional study of agential design within the larger focus of user-centered design and methods that seek to bring to light users' conceptions of agency during navigation. Recognizing how users feel about the shifts in agency for various navigation-related tasks could help technical communicators convey whether users are confused or antagonistic to new autonomous features, and whether users are able to incorporate those features into their own user *techne*. We can also seek out methods to identify how agency is distributed during navigation, moving beyond the user's perception of agency. The problem with locating agency – in any context, including autonomous or semi-autonomous vehicles – is that any method we employ to understand agency will need to be somewhat abstract, or run the risk of distorting the representation of agency. The following hypothetical example showcases these problems.

In thinking of the complex, immersive environment that I have been calling a user *techne*, or a *techne* of navigation, it might seem reasonable to move back to a concept I pointed out in Chapter 2 as key to the study of persuasive movement – *kinesis*. We may be tempted to consider whether the study of *kinesis* movements would aid us in identifying agency. For example, although it would take a long-term observation of the driver-vehicle environment and perhaps a few dozen engineers, it *would* be possible to record most of the *kinesis* movements that occur during navigation. This observation of movement would not be limited to those of the human driver but would include those of nonhuman agents as well. We could count the position of other drivers on the road, the movement of the steering wheel, the movement of the pistons, and the automatic setting and adjustment of the speed by a smart cruise control system. In order to fully account for all of the actants, we might be tempted to create a large table of their varying relations and powers and chart all of the movements they perform or enable to be performed during navigation. Thankfully, such a task would be both maddening and unnecessary, not only because no one would be interested in doing the work but also because even a complete table of agential movements would not "account" for agency. Examining *kinesis* movements could aid in the identification of actors, though *no list of actors and the movements they perform will fully account for agency*. For a more useful account of agency, we would also want a rich description of how the most influential actors interact. Rich descriptions of these exchanges could illustrate the flow of agency among these actors as well as the movements they perform. Yet, since agency cannot be possessed but embodied and performed, researchers utilizing these concepts could map actors' roles so that they form a conversation representative of agency, much as Bruno Latour does, albeit with alternate methodological assumptions, throughout the book *Aramis*.

In *Aramis*, Latour employs his "ontosociology" and identifies the actors and the forces that make up a sociotechnical situation. Latour focuses on the *network of actors*, and his methodology is to scrutinize those in the network of forces that impact a particular question of agency. He follows both humans and nonhumans, because, as he argues elsewhere, "[n]o observer of human collectives, for at least the past two million years, has ever been faced with pure social relation, and none of course, *especially* in high-tech modern settings has ever been faced with a pure technique" (1993b, 380). In order to identify all of the translations at work within a network, Latour moves slowly, pausing to listen to every *thing*, because his guiding focus is to avoid discriminating between humans and nonhuman agents. Thus, one of his primary maxims is to consider all sources and "keep everything flat" (2007, 190). Latour's methodology – and his representation of that methodology in *Aramis* – could provide technical communicators with a template for exploring complex, situated realms of agency, especially in tracking agency between humans and nonhumans and identifying how things in the world perform the work of translation. The advantages to attempting such a Latourian study while attending to *kinesis* movements are threefold: we identify the differences between agent and agency, we become more attune to all actors' function as translators, and we avoid locking nonhumans out of agentive roles – the roles they so obviously play in everyday life.

My goal here is not necessarily to identify which methods will be applicable for transportation projects, but to prompt further questions about the connections between usability testing and user-centered design, considering the complex agency of automobile use. What we might call "agential design" needs to borrow from existing rhetorical theory and mix that theory with usability methods. While these studies may not always fixate on the human user, we can still advocate for users throughout the design process, identifying what will make them happiest, safest, and more accepting of autonomous technologies in transportation. In order to attain these goals in transportation design, we will likely confront the issue as to how much of the "bit" humans will want during navigation. We might think of this research (and hopefully this design) as user-centered, but agentially distributed.

Agency and the role of the "bit" plays a strong role in transportation, and this chapter has shown that transportation designs have sought ways to minimize human errors, from "sleeping at the wheel" or misjudging the movement of other drivers in the immediacy of vehicular navigation. I have also advocated for the complexity of use, showing that use is a *techne*, like design, and has always involved agency distributed across many actants, from chariot driving and ship steering to modern-day vehicle navigation. Recent rhetorical scholarship on *techne* aligns with classical Greek articulations of a *techne* more concerned with the using arts than making arts. From myths about Athena's role in both design and use, to the language of carpentry and ship navigation, there exists a user-oriented *techne* alongside a making-oriented *techne*. Further, this *in situ* use connects to rhetorical conceptions of *metis*, *tyche*, and *kairos*, as well as with driver-vehicle environment studies. In the next chapter, I explore one specific post-automobility project. While this project failed and the vehicles never saw public use, the case study allows us to examine how one transportation project was ultimately unconvincing and did not successfully employ a persuasive kinesthetic rhetoric.

Notes

1 Toyota's 2010 recall of sticking accelerator pedals did not include the Prius.

2 A case can be made for medicine, military combat, and others as well. I should also make it clear that we can also refer to a user *techne* as a *techne* of "practice," since some arts, like medicine, military combat, and navigation, certainly are *practiced*. I prefer the term "use," however, because it: 1) balances out the concept of a "making" *techne*, 2) points to specific technological manipulations, and 3) suggests that any "user," whether experienced or novice, engages in a *techne*.

3 Aristotle uses this phrase multiple times in Book Two of *Poetics*.

4 The first reference to *kairos* comes from Hesiod's *Works and Days* where it appears alongside advice to a worker not to overload a cart, lest the weight break an axle.

5 For a more complete elaboration on *kairos* in rhetoric scholarship, see Debra Hawhee 2005, pages 66–71.

6 I have previously explored how alphanumeric discourse can mediate the embodied "feeling" of driving a car, though this study concerned gender and motorsport (Pflugfelder 2009).

7 Another study exploring agency and driving, "Is No One at the Wheel? Nonhuman Agency and Agentive Movement," has been published in a book-length collection on rhetoric and Bruno Latour.

7 EPTV

The following case study relates an innovative transportation design that I was able to study up-close for a period of two years. This design project, under development at a large Midwestern land-grant university from 2007 until 2011, I will call the Electric Personal Transportation Vehicle project, or EPTV. The project involved the design and manufacture of an all-electric, self-guided (autonomous) transportation system centered on the design and production of a small vehicle known as a "pod car." Ultimately, dozens of these pod cars were to be built and eventually connected to an information and guidance network. The network was never finished, but we can now examine the design of the vehicle to understand why it was not persuasive – as a design project, as a site for new interfaces, as a new logistic, and as a new *techne* of navigation. Before I focus on this analysis, however, I explain the context of the design project and then consider the formation of EPTV through the lens of actor–network theory, employing a similar framework from Bruno Latour's post-mortem analysis of Aramis in order to do so.

The EPTV project

The concept for the EPTV came from a professor in the school of mechanical engineering who regularly taught a project design and development course. In this course, which is a requirement for all seniors, students are typically asked to work in groups on a semester-long project of their choice. These projects often involve the design and manufacture of a specific mechanical object. For the 2007 version of this course, instead of giving students free-reign to develop their own projects, the professor pursued a unique thought experiment: what could transportation at this university look like in the coming decade, assuming that the school implemented the construction projects outlined in the most recent campus master plan. Architects and administrators wrote the master plan with the assumption that the main roads that run through the campus would be changed into light-duty pathways and heavy traffic would be diverted around the school. Students in the course were asked to design an environmentally sustainable transportation solution able to function within a large central campus that could not accommodate a slowly increasing volume of buses and automobiles. The solution that the course collectively developed became known as the EPTV personal transportation vehicle (see Figs. 7.1 and 7.2 for CAD renderings of the pod car).

As a technical project, EPTV was an interesting engineering design exercise, integrating several emerging concepts in transportation design. First among the EPTV's notable technical features is its autonomous navigation system. The vehicles were guided by three technologies: GPS-guided software that sends messages to the steering, throttle, and brakes; dead reckoning, which estimates distance and direction; and, eventually, a laser scanner that keeps the PTV (personal transportation vehicle) on a predetermined path marked with specific visual cues.[1] The PTVs did not require rails or tracks and were designed to be as flexible as possible with the minimum amount of coordinated infrastructure. Laying rail lines would cost many millions and disrupt traffic flow inside the campus as much as, or more than, the streets they were to replace. PTVs could be reprogrammed and their routes adjusted in an ad hoc way so that the entire network could be altered day-to-day or hour-to-hour. The PTVs were powered by an electric motor with replaceable banks of batteries, designed to run for eight hours before being recharged. A few similar projects have been attempted in the past. West Virginia University's rail-based personal rapid transit (PRT) system is one of the more famous examples (see Fig. 7.3), though it does not function with an on-demand capacity. There are now PRTs in Heathrow airport and Suncheon, South Korea, and Abu Dhabi's Masdar City is planning for a similar

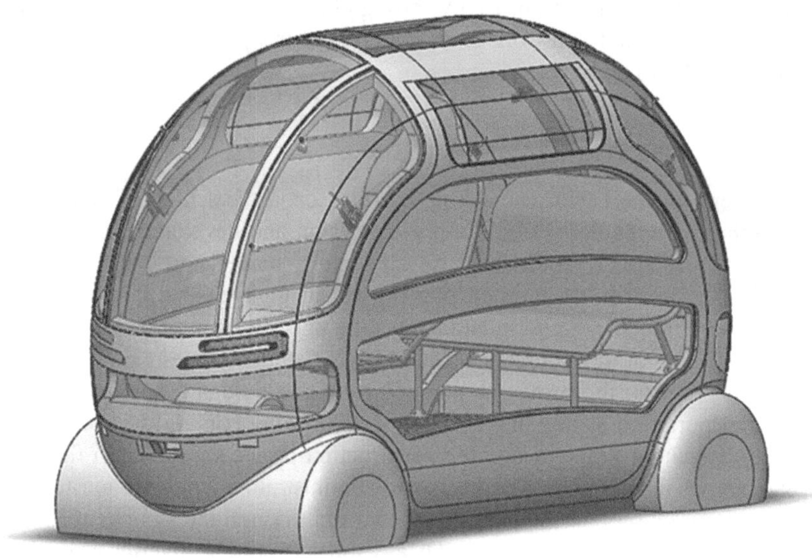

Figure 7.1 A rendering of the EPTV pod car bodyshell as it was being developed as a proof-of-concept in spring, 2010

Source: Author and EPTV Club

Figure 7.2 A rendering of the PTV chassis (facing away)

Source: Author and EPTV Club

Figure 7.3 West Virginia University's PRT

Source: Marshall 2008

Figure 7.4 Masdar City's PRT

Source: Seifert 2012

system (Fig. 7.4). A very similar system, which debuted in 2015, is the autonomous Wepod network, set up to transfer people between the towns of Ede and Wageningen, in the Netherlands.

Unlike some of the larger rail-based PRTs, EPTV pod cars were initially designed to have a two-person capacity (later changed to four) and were estimated to weigh roughly 3,000 lbs., though the weight steadily increased throughout the project. EPTV cars included a center drain, a touch-screen interface, ambient lighting, collision-detection sensors, four-wheel steering (initially), and a number of on-board cameras (eventually). The design for the door changed several times, from sliding, to gull-wing, and finally to a multi-articulation setup. EPTV vehicles would not operate on city streets, as they would not be crash-tested, and they could not achieve roadway speeds, as they were limited to less than 20 mph. In order to board an EPTV pod car, riders would text the EPTV system with their cell phone and the closest vehicle to the designated stop, moving in the same general direction, would deviate from its original route in order to pick up the rider. While none of the particular technical features planned for the EPTV is new to transportation technology, their assembled integration made the undertaking rather complex for an undergraduate engineering project. The proposed rollout and promotion of the EPTV was also ambitious, as the club was told that PTV pod cars would be showcased during the 2012 Super Bowl.

After several semesters as a course-based technical project, several students along with the initiating professor created the EPTV Club. This club was officially recognized by the university, had weekly meetings, and was dedicated to manufacturing a pod car and developing the EPTV network. The club included an executive board made up of 11 members, one main advisor (the same founding professor) with an attendant EPTV advisory committee and a number of working groups. These groups focused upon specific technical projects within different engineering domains, with the larger hope that smaller teams of three to seven students would be more efficient in making technical decisions and carrying out production.[2] These groups were also broken down into smaller teams with specific responsibilities. For example, the Controls Working Team was sub-divided into Operations (OPS), Obstacle Detection and Avoidance (ODA), and Navigation (NAV) foci. Though initially made up of broader "Logistics," "Engineering," and "Manufacturing" working groups, eventually the club formed six different teams:

- Controls: planned and formed basic controls systems
- Endo: developed the interior
- Exo: designed iterations of the exoskeleton
- Manufacturing: assisted other groups with build processes
- Navigation: coded and wired the navigation systems
- Marketing: promoted and branded the technical project

Students who identified as technical communicators were not involved in the preliminary stages of the project, but eventually found roles within the club, most often in the "Marketing" team. From the fall of 2009 up to spring of 2011, I held the role as the lead technical communicator in the EPTV Club, the head of the Marketing working group, and a member of the executive board.

EPTV's decline was rather sudden – and we cannot attribute the failure to a lack of student interest, funds, ambition, or oversight. As of the spring of 2010, the EPTV club included several dozen active members, not including students involved in the project from their mechanical engineering coursework (several dozen more). Further, though I have not been able to pinpoint a specific monetary figure for the cost of the EPTV project, the cost for the materials to build one car was estimated at roughly $27,000. Since most of the parts for the pod car were purchased or fabricated, we can assume that at least this total was spent during the course of the project. The initial concept for EPTV was also vastly more complex and expensive than I have represented it thus far. The yearly maintenance cost of the fleet of pod cars was estimated at $100,000 and yearly electricity and repair estimates multiplied this sum many times over. The ten-year estimates for the cost of running EPTV ran into the tens of millions. That said, the EPTV project never produced the desired number of pod cars, built the charging stations, programmed the software, or found room for EPTV storage, maintenance, and charging stations.

Despite the momentum created through all of the planning, design, and engineering work, the EPTV project sputtered to a halt in the spring of 2011. The club's members had mostly left, meetings were no longer scheduled, and the manufactured materials, including the proof-of-concept EPTV pod car, were dismantled and re-purposed in other mechanical engineering projects. So, what had happened to the inertia generated in the senior design course and carried through into the student club? How did a well-funded, enthusiastically supported project that involved dozens of people cease to exist over the course of approximately four months? Below, I try to understand what would have made EPTV successful, and how technical communicators could assist future transportation projects and avoid the issues that EPTV faced. The methodology upon which I ask these questions is informed by other actor–network theory research on failed transportation projects, and the concepts from a kinesthetic rhetoric that have been central to previous chapters.

Failed transportation projects

Understanding why transportation projects fail is a task that has been taken up in science and technology studies by researchers who employ sociological approaches to technology in order to make sense of how projects fall apart. Often, the demise of a city-wide, multi-million dollar project calls for an approach that abstracts the major players involved, so as to not blame one part or technology for why an entire system failed and thus fall into a kind of tautological or technological determinism. Aside from studies of catastrophic failures, such the research on the space shuttle Challenger disaster, technical communication researchers have not tackled too many large-scale failures.[3] In part, this may be because most research reports on successful applications of technical communication know-how, with the goal of presenting advances in pedagogy or offering case studies that can be further developed. Studies from within social construction of technology, science studies, actor–network theory, or related fields or methodologies, however, have presented such failures as ways to understand how sociotechnical networks are built and maintained, or alternatively, how they disintegrate. These studies provide us with a much broader view of the networks that need to be developed and sustained for such projects to be successful.

Research on failed transportation projects from the perspective of actor–network theory include Michel Callon's "The Sociology of an Actor-Network: The Case of the Electric Vehicle" (1986), John Law and Callon's "The Life and Death of an Aircraft: A Network Analysis of Technical Change" (1992), and Bruno Latour's *Aramis, or, the Love of Technology* (1996). Callon's study sought to understand the *Electricité de France's* (EDF) efforts to develop an electric car along with the *Compagnie Générale d'Electricité* (CGE) and automaker Renault. In his analysis, Callon posits that no technical object can be understood without considering its concomitant actor-world, where any project is only as durable as the associations of the heterogeneous elements in that world. Such associations are built out of the work of translators, the actants that define roles and scenarios, and the process of "simplification," where complex institutions, spaces, and technologies are reduced to their most important relations. In the end, Callon identified

Renault's desire to create alternate actor-worlds and competing translations as a main cause of the project's failure. Law and Callon later investigated a British aerospace project from the 1950s and 1960s – the creation of a tactical strike and reconnaissance plane for the Royal Air Force called the TSR.2. Building from earlier work in ANT, they argued that successful projects engage a global network that provides resources, but does not micromanage, and a local network that provides a product, but also leaves room for experimentation. They found that the TSR.2 failed because there was too much reciprocal over-simplification, where complex institutions with many divergent opinions and functions were simplified by each other, resulting in a skewed representation. Law and Callon also identified: too many changes in the global network, an inability to build a satisfactory local network, mistrust between actors in both the global and local networks, and "seepage" between the two, as both involved actants trying to influence the scope of the other.

Distinct from Law and Callon's work, Latour's methodology in *Aramis* is in no way a straightforward concern. The *Agencement en Rames Automatisées de Modules Indépendants dans les Stations* (Aramis) was an urban transportation project developed by the French government, the government agency *Délégation Interministérielle à L'aménagement du Territoire et à L'attractivité Régionale* (DATAR), the *Régie Autonome des Transports Parisiens* (RATP), and the company Mécanique Aviation Traction (Matra), among other institutions. From 1969 to 1987, thousands of engineers, designers, officials, and managers worked to complete Aramis, which was to replace the Paris metro. Basically, Aramis was a light rail system, but with some major differences. Instead of large rail cars, Aramis used small rail cars capable of holding only a few people and did not link these rail cars together with direct, mechanical links, like a traditional train, but with what were called "nonmaterial couplings." These couplings also gave Aramis its main advantage – riders could tell the system where they wanted to go and one rail car could "pull away" from an unconnected chain of cars and pick up a single passenger. With this method, Aramis could operate like a network of independent, yet still linked, rail cars, each focusing on taking small numbers of people to their destinations. For several reasons, and after a number of side projects and test runs (see Fig. 7.5), the project was halted in the mid-eighties, after 18 years and the expense of millions of Francs.

Latour explores why Aramis was not implemented through the fictional interaction of two main characters, a sociology professor who has been tasked with writing a report as to why Aramis "failed," and a sociology student learning the professor's unorthodox methodology. A rough analog for Latour himself, the professor employs a kind of "relativist" sociology, where groups of influencing actors, human and nonhuman, are considered equal. Latour focuses on the *network of actors*, and his methodology is expressed as a drive to "follow the actors" in what they do and where they go (1996, 10). To represent his focus on every influencing actor, from ministers of government and engineers to technologies, laws, and protocols, Latour includes transcripts of interviews, official documents, and even the fictionalized voice of Aramis itself.

Figure 7.5 Aramis test track by the Gare de Boulevard Victor, Paris
Source: Smiley.toerist 2010

For Latour, no technological project is doomed from the start any more than any person is confined to a life of failure after he draws his first breath. Aramis was as likely to succeed as the next autonomous transportation project. Claiming otherwise is technological determinism – where successful projects succeed because they are better and more advanced. As Latour argues about the Aramis project, "[y]ou aren't born feasible or infeasible; you become so" (122). Latour dismisses any "talk of inherent flaws" as a "tautology. It's immoral; it's kicking a thing when it's down" (120). No project has its life traced out from its conception, but functions within a network of actors that impact its trajectory. Commenting on PRTs in general, Latour argues:

> You can't say that PRTs died because they weren't viable, any more than you can say that dinosaurs, after surviving for millions of years, died out because they were doomed or ill-conceived. Aramis died – in 1987 – and its accusers claim that it was nonviable from the beginning, from 1970. [. . .] No, Aramis is feasible, at least as feasible as dinosaurs, for life is a state of uncertainty and risk, of fragile adaptation to a past and present environment that the future cannot judge.
>
> (35)

To avoid a determinist perspective, he develops a methodology that *cannot accuse in retrospect*. Using this logic, Latour identifies why Aramis failed as a large-scale transportation project.

Latour claims that Aramis was shuttered for three main reasons. First, while typical project-planning occurs in identifiable phases, Aramis, as a deeply innovative technical project, did not. Latour explains that the traditional move from "less real" to "more real" in project management was not necessarily a problem, but Aramis became too confusing for design teams to follow. The character of "Mssr. Laredo," a specialist in science policy and the evaluation of research programs, argued that some Aramis project staff "were going back to Phase 1, to basic research, and at the same time others were settling the question of the soap the maintenance staff was supposed to use to wash the windows" (271–272). Certainly, there were some project-planning issues. Second, Aramis never absorbed any of the contradictions of its concomitant networks. Those involved in the project believed in the initial concept and never wanted to see it change over time. The result was that Aramis remained "purely an object, a pure object" (280). Perhaps, Latour speculates, "[t]hey abandoned Aramis so as not to compromise it" (288). Aramis was never given the chance to be affected by the huge number of actors within its network and because of this isolation, it remained close to the initial concept until almost the very end. Finally, Latour argues that those that worked on Aramis did not love it enough, though by "love" Latour means that no one cared enough about Aramis to passionately research it. The sociologist professor accuses Aramis officials with a scolding tone: "Oh, you do love science! [. . .] But you don't love research. Its uncertainties, its whirlwinds, its mixed character, its setbacks, its negotiations, its compromises" (293). Here, "love" is representative of a passion for the Aramis concept and the kind of continuing interest and adaptation that researchers should have pursued. According to the "voice" of Aramis in Latour's text, engineers in the project never fought over the details. Aramis itself states, "you humans need to talk, argue, get mad, that's your role in this imperfect world," though they did not seem to do so (295). This conciliation arose because those involved in the project saw Aramis as a utopian ideal – with it, the transportation problems of Paris would be solved. Aramis again lashes out: "No, no, you didn't love me. You loved me as an idea. You loved me as long as I was vague" (294). Such language might be surprising, especially coming from the voice of a transportation technology, however, Latour's concept of a lack of "love" means something else entirely if we consider the kinesthetic rhetoric developed in previous chapters.

Understanding EPTV

In order to understand the EPTV project, I borrow a number of Latour's assumptions about actor–networks and about the plausibility of large technological projects, while also deploying concepts from the kinesthetic rhetoric explained in previous chapters. Like Latour, I assume that actor–networks are assembled by powerful agents and that those same agents bring a wide array of actants, both human and nonhuman, into relations. I also assume that all technological projects are possible and that no project is doomed from the start, though these projects need to absorb internal contradictions that will form as the network changes. Any transportation project can exist if it manages to accommodate many definitions, change with these accommodations, "hold its whole contradictory environment together," and eventually make the contradictions disappear (Latour 1996, 207). Then, and only then, can a technological project be seen as an object for use. For example, EPTV would have succeeded if and when passengers began saying, "I'm going to the store," without mentioning how they would get there, with a tacit understanding that the EPTV would be used. They would take the EPTV, as it were, for granted. Finally, Latour argues that technological projects need to be "loved," though his conception of "love" is unsatisfying. In its place, I consider whether EPTV, as a technical project, was sufficiently persuasive.

To understand why EPTV failed, I employ four specific methods. First, I use a timeline to show how the EPTV project was able to address problems that arose during the design process. Building from the timeline, I then chart out the major actants that impacted EPTV. Using these actor–network diagrams, I look to identify which connections impacted EPTV the most and how impactful those relationships were. Third, I identify as many definitions of the EPTV project as possible, using meeting notes and data from a series of interviews.[4] With this method, the goal is to see how many contradictory definitions emerged throughout the EPTV project and whether or not those many definitions were reconciled. Finally, I consider EPTV from the perspective of a kinesthetic rhetoric, evaluating EPTV from the standpoint of design, interfaces, logistics, and navigation.

Charting EPTV in a timeline assists in identifying some of the many actors involved in the EPTV network, including technologies, materials, club members, and institutions, so that these actors may be later charted. A rough timeline for the engineering design and project-planning aspects of the EPTV project is as follows:

- Fall, 2007 through summer, 2008: The EPTV design is conceived as an engineering project for students in a senior-level mechanical engineering course. Students, under the guidance of a professor, develop a complex, pod car-based transportation system and design a number of features, including the logistics, chassis, drivetrain (the electric motor, power supply, charging system, and gearbox), steering, braking systems, suspension, and controls. A cardboard mock-up of the entire pod car is developed.
- Fall, 2008: EPTV is refined through a number of different design reviews, as specific parts of the pod car are redesigned with CAD software. A 1/12 scale model of the pod car is also completed.
- Spring, 2009: EPTV undergoes further specification changes, as students build and design more components of the pod car. Students decide on the complete assembly specifications of the vehicle, though not the exterior design, and render everything in SolidWorks CAD software. Some of the "skateboard" (the chassis for the pod car) and the electronics are fabricated this semester. Design changes occur to the door assembly, the wheel size, and the body shape. Obstacle detection and avoidance becomes a concern and students design a bumper with collision-detection sensors. A 2004 Yamaha

Rhino 660 ATV is purchased to source suspension, brakes, and wheels. The EPTV Club is formed during this semester with the goal of working on the EPTV project outside of class time. A shared network drive space is started for storing build details. Students not in the senior design course are encouraged to join and gain hands-on engineering and manufacturing experience.

- Summer, 2009: Larger portions of the EPTV skateboard are completed, including a union of suspension, braking, drivetrain, electrical, and battery components. Electrical components are simplified and ordered, and a chassis is completed. The overall design undergoes further revision, as the exterior door is built, passenger capacity increases from two to four, the touch screen moves from the end of the vehicle to the side, and new seats are designed. A bumper design is put on hold. Additionally, an industrial design course is enrolled in the EPTV project and students design a number of different iterations of the EPTV exterior and a slogan: "Campus Vehicle for Next Decades" (sic).
- Fall, 2009: The fall of 2009 brought greater involvement in the EPTV Club from students outside the mechanical engineering course and further refinements to the components and drivetrain. Many of the components came together during this semester, resulting in a powered, if not thoroughly driveable, pod car chassis. The EPTV Club creates a Website, generates a logo, and begins using Vault, a version control software (VCS).
- Spring, 2010: There is no mechanical engineering course involvement this semester, but club work continues. The club's organization is restructured, there are new attempts at project documentation, body components are fabricated, a new door is designed from wood, lighting fixtures are purchased, and the suspension and steering are stabilized. The pod car is driveable and steerable – small-scale movements of the vehicle are performed.
- Summer, 2010: No work occurs.
- Fall, 2010: Coding issues with navigation are solved, battery recharging issues are solved, and plans are developed for testing the GPS code and directional controls on a modified radio-controlled car.
- Spring, 2011: Few EPTV Club members attend meetings and the EPTV pod car chassis and EPTV Club computer are moved to a new room. By the end of semester, the EPTV chassis is dismantled and parts are distributed into bins to be used in other mechanical engineering projects.

We can see that while the EPTV project gained momentum throughout the project, the final year showed signs of slowdown. There were still significant numbers of EPTV club members through the fall of 2010, but by the spring of the following year, the club and the project were in serious doubt. However, if we consider the basic relationships generated by the EPTV project, there are no obvious signs of contention. The school of mechanical engineering provided financial support, advisors gave guidance, and progress on the design and manufacture of the pod car continued at a reasonable pace. Examining the life of EPTV chronologically does not point to any specific actors that led to the failure of the project, though this analysis has identified the major phases of the project.

Actor–network diagramming

Though Latour found it unnecessary to chart out Aramis' actor–network through diagrams, they prove useful in evaluating some of the problems that occurred with the EPTV project. These diagrams can represent the strength of different ties within the network, in order to identify when particular actants gained or lost influence in the project. In the system I use below, circles represent different actants, and stronger, bolder lines represent more powerful relationships. (Actor–network diagramming is not an exact science, so these lines are not intended to have a specific mathematical equivalent.) In looking at when EPTV was emerging from a course project and the EPTV Club began (Fig. 7.6), there are strong connections between the two, because much of the information generated from the course and represented in end-of-semester critical design reviews were also incorporated in the club's work. Here, we can also see that the pod car itself was the main focus of attention for the club.

When we look at the EPTV project about a year later, when the club was going strong and when senior design courses stopped working on the EPTV project, we recognize that there are many more actors involved with the project. In Fig 7.7, we can see considerable focus on the production of the pod car, which the club

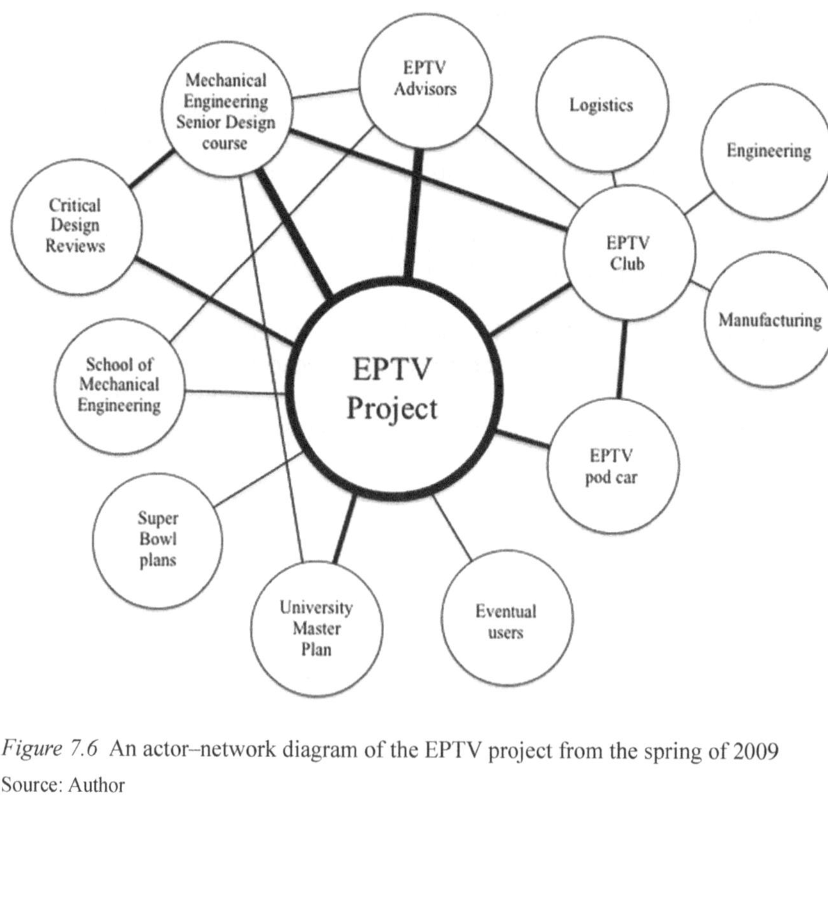

Figure 7.6 An actor–network diagram of the EPTV project from the spring of 2009

Source: Author

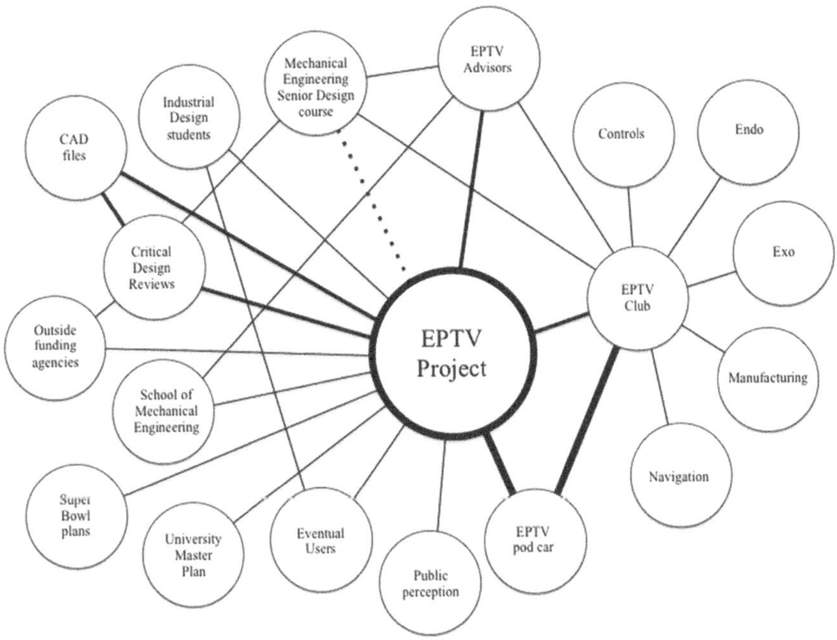

Figure 7.7 An actor–network diagram of the EPTV project from the spring of 2010

Source: Author

was developing with materials created from the courses and their critical design reviews. The club is more dependent upon CAD files in particular, some of which had come from the course, and the club has grown larger in scope, with more students and more working groups. Also appearing at this point are greater considerations of the eventual public users of EPTV, as identified by a class of industrial design students working on the project. Plans to show the EPTV pod cars at the Super Bowl have receded, as has the focus on the university's master plan.

Near the end of its life, we notice some new trends in the EPTV actor–network (Fig. 7.8). While users (and the broader public perception of EPTV) are still in the picture, they are pushed toward the edges of active consideration. Likewise, the primary impetus for the EPTV project, the university master plan, has receded into the background, along with any goal of displaying the EPTV pod car at the Super Bowl. The loss of these long-term project goals could either be the cause or effect of other problems in EPTV. While we see the addition of the marketing working group, the development of a remote control version of EPTV (which was to be a temporary public face of EPTV), and the introduction of version control software (VCS), these new actants did not seem to have significant impact upon the project, even though the introduction of VCS attempted to bring systematic order to the use of critical design review-guided CAD files. AT this point, we could describe the project as being dominated by consistent, grounded work to

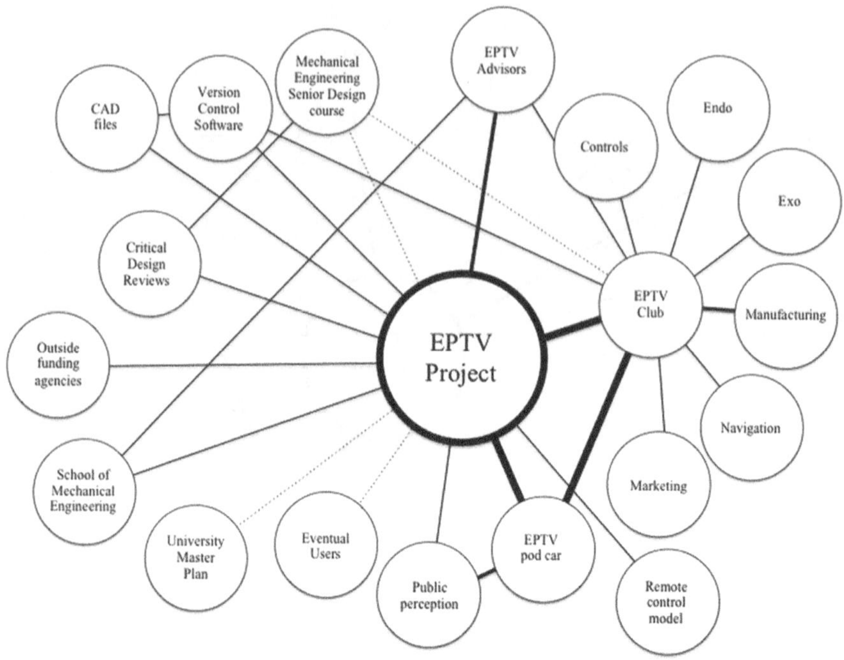

Figure 7.8 An actor–network diagram of the EPTV project from the spring of 2011

Source: Author

build the pod car, with much of this work being accomplished from a mechanical engineering perspective.

With these actor–network diagrams, we can argue that while the EPTV project appeared to be supported by a robust network, the particular effectiveness of the strongest actors in that network is suspect. EPTV certainly enrolled significant numbers of people and the mechanical problems with the pod car were being resolved as the project developed. Over several years, power in the network shifted from the pedagogical goals of the course to the club's manufacture of the pod car. Perhaps this focus was too narrow, the transition from the course to the club problematic, or the focus on the production of the pod car itself limited in scope. These are several possible weakness in the project. Yet, the problems with EPTV are not solely available through an articulation of the strength of actant relationships. We also need to look into the kinds of assumptions that developed from them. Perhaps these definitions fragmented the project and contributed to EPTV's failure to cohere as a stable transportation concept. We can perform another common ANT method in order to examine these changing definitions.

Defining EPTV

Judging by the timeline, the EPTV project was progressing with momentum when nearly all interest in the club ceased. Judging by the EPTV's actor–network, there were always a number of actors engaging with the EPTV project, supporting it, and thus defining it. So, what caused the EPTV project to grind to a halt so soon after the pod car sputtered to life? Perhaps even before we can begin to tackle the question of how EPTV died – for there was no murderer, but a death certainly occurred – we must first ask *what EPTV was*. The banal answer is that EPTV was what I have been calling it all along in this chapter: an innovative transportation project, born from an engineering problem, and supported by a mechanical engineering department at large university. But EPTV was more than that, much in the same way that Latour's Aramis was more than an innovative rail car project supported by a combination of public and private funds. EPTV was a series of different definitions, because it meant many different things to the many different actors that comprised its network. Each student engineer, faculty member, department representative, researcher, and potential rider has an equally valid perspective as to what EPTV is. Latour notes the same for Aramis, arguing that, "[t]here are as many possible Aramises as there are minds and points of view" (1996, 77). Yet, in order to succeed, EPTV had to absorb those contradictions.

So, just how many EPTVs were there? The following list is a rough approximation of some of the definitions of EPTV. These definitions are based upon the critical design reviews, my meeting notes, and the interviews I conducted with EPTV club members.

1. EPTV is a pedagogical device within a mechanical engineering course that engages students in engineering design and manufacture (EPTV was initially developed for a mechanical engineering course).
2. EPTV is a reaction to the 2020 master plan (EPTV was developed as a thought experiment that emerged from a university's master plan).
3. EPTV is a complex network of instructions and algorithms designed to optimize movement efficiency (EPTV's autonomy requires a coordinating computing network).
4. EPTV is spotlight for a research university's engineering prowess (EPTV is often shown to tour groups and potential students).
5. EPTV is the hard work of mechanical engineering students fulfilling the requirements of a mechanical engineering senior design course (EPTV is something students work on for experience and acts as a means for evaluation).
6. EPTV is the focus of a university club, with 45+ members, whose aim it is to advance the project while educating students (EPTV transitioned from a course project to a club project).
7. EPTV is a self-locating, self-driving, highly responsive mobility technology designed to become the default transportation solution for students, faculty, and staff (EPTV is intended as public transportation).
8. EPTV is a collection of parts, some built specifically for their purpose, some extracted from a disassembled 2004 Yamaha Rhino 660 ATV (EPTV is a mechanical engineering fabrication project).
9. EPTV is an evolving 3-D graphical representation of a series of engineering constraints, manipulated by CAD software (EPTV is designed in CAD software before being manufactured).

10 EPTV is a representation of the post-car world and futurism in popular culture (EPTV is the promise of progressive thinking).
11 EPTV is a series of critical design reviews, in PowerPoint form, and the analysis of those slideshows by EPTV Club members (EPTV's progress is articulated through students' and club members' design reviews).
12 EPTV is a safety device (EPTV includes a wide array of safety features and is intended to be an extremely safe mode of local transportation).
13 EPTV is small, autonomous pod car designed to carry four passengers between destinations on campus (EPTV is a short-distance, convenient, university-bound mobility device).
14 EPTV is a green technology, representing a commitment to lower carbon emissions (EPTV is an all-electric extension of the university's desire to have a smaller carbon footprint).
15 EPTV is a node on a network, representing the connections between a nearly infinite variety of trajectories (EPTV pod cars are not bound to a fixed pattern – its pathways are highly mutable).
16 EPTV is an engineering exercise that shows the level of hard work and planning that takes place at a state-funded engineering-focused university (EPTV is represented as a student project at events and conferences and shown to donors and state representatives).
17 EPTV is collection of end-of-semester reports and Gantt charts that detail the activities of EPTV Club's working groups (EPTV's progress is articulated by the club through different documents).
18 EPTV is an array of individual projects, without a strong design coherence, completed for a mechanical engineering course (EPTV's coherence is often difficult to understand by the club).
19 EPTV is a far-fetched idea that cannot be accomplished in the manner originally defined and with the initial goals in place (EPTV is the subject of skepticism from the wider public who have not seen the progress of the project).
20 EPTV is a compromised version of the original designs and an eventual compromise to existing time and resources (EPTV necessarily adheres to the constraints of student schedules and the resources made available by the school).
21 EPTV is a proof-of-concept of an autonomous pod car (EPTV's pod car is not a prototype of the completed product, but constitutes working proof that the project can be completed).
22 EPTV is a drivable frame with a wooden and fiberglass bodyshell (EPTV's pod car can be driven around campus).
23 EPTV is a mobility assistance device focused on the needs of the handicapped members of the university's population (EPTV becomes briefly re-articulated as an accessibility device as the club seeks more momentum).
24 EPTV is a radio-controlled car that stands in for the larger pod car (EPTV's guidance systems are tested out on a radio-controlled car).
25 EPTV is an old engineering project pushed to the side to make room for more current projects (EPTV's pod car is moved out of the way when mechanical engineering students need the room).

26 EPTV is different parts that can be used in other projects (EPTV's pod car is dismantled and sorted into parts bins in the basement of the mechanical engineering building).

Did EPTV absorb these contradictory definitions? Certainly, the final definitions point to its existing downfall as a project. Earlier definitions, however, do not illustrate the precise difference between the school's understanding of EPTV and the club's understanding of the project. That said, any innovative transportation project that includes a history involving so many definitions is likely to have difficulty merging them into a coherent whole. Some of the more significant contradictions appear in the gap between definitions that identify EPTV as: existing transportation and an on-going project; a public project and an engineering education exercise; a material pod car and a series of software-readable files; and an easily realizable design and a group of related problems that require constant attention. There are likely more contradictions, though these stand out as significant to the success of the project, as they do not appear to be contradictions that were resolved.

So, after looking at EPTV from the perspective of its timeline, its actor–network, and its multiple definitions, we now know that it did not have an obvious and direct downfall, though the EPTV Club paid more attention toward the production of an individual proof-of-concept pod car while some of the larger goals for the club shifted out of focus. We also know that EPTV involved dozens of possible definitions, as the project changed over time, some of which likely caused problems in the unity of the project. While it may be easy to suggest that EPTV would have never succeeded as a transportation project and would never be completed, if we dismiss technological projects only because of their so-called "feasibility," we run the risk of not understanding the complex issues involved in their undertaking.[5] Latour has argued that Aramis, among other things, was not "loved" enough, though this term means something else in the context of a kinesthetic rhetoric.

In Latour's work, Aramis speaks and tells the engineers and administration officials "No, no, you didn't love me. You loved me as an idea. You loved me as long as I was vague" (294). Here, love stands in for a number of concepts. By love, Latour really means that Aramis was not allowed to change to meet the needs of the actors in its network, Aramis was only appreciated as an abstract concept and not a changeable agent, the project could not corral the many definitions it spawned, and its designers never argued passionately about its success. In short, Latour believes they did not care about Aramis as a project that needed attention and guidance, but prematurely objectified it. It was not nurtured through its development, but expected to become great on its own. For Latour, that perspective is shorthand for a lack of love. While I do not doubt Latour's claims about Aramis, I suggest that a lack of love is not a particularly good representation of what happens when projects fail, Aramis or EPTV. In order to explore an alternate account of EPTV's death, I examine the project from the perspective of the four major concerns of a kinesthetic rhetoric: design, interfaces, logistics, and navigation.

A kinesthetic rhetoric suggests that EPTV failed because it was *unsustainably compelling* to designers and users. Further, there was a significant gap between the three conceptions of *techne* generated during the project.

Design

The design of the EPTV project was carried out by two student populations at two different phases. First, mechanical engineering students in a senior design course working under the guidance of a mechanical engineering professor completed several semester-long projects. These groups chose to focus on (what they believed to be) discrete aspects of EPTV, as their grade was dependent upon designing and completing a focused project within the 15-week course. They tackled specific components such as the drivetrain, the electrical system, and the suspension, and created project plans, kept up-to-date bills of materials, designed parts with CAD software, and then manufactured or assembled as necessary. At the end of the semester, students completed a critical design review, presented in Microsoft PowerPoint, and showed that their finished productions worked, given the relevant requirements and constraints. When the EPTV Club was formed, club members, who were also students, relied upon these design reviews and chose to keep many of the work divisions first established in the course. Further, the EPTV Club's design process emerged from club members' major fields of study. That is, the processes of invention, design, fabrication, and assembly were dominated by the field of mechanical engineering (ME). This comes as no surprise, as many members of the club were mechanical engineering students and the project came out of and was housed in the mechanical engineering department. In interviews I conducted with club members, more than one student explained that "ME-thinking" enveloped the entire project, influencing the trajectory of the design – for example, that the vehicle frame was built first and conceptualized as an entirely unique fabrication (i.e., it was not built upon an existing frame). This art of design was largely focused on the week-to-week creation of specific, realizable materials and design goals, and can thus be characterized as making-oriented *techne* that concerned know-how and realizable outcomes. In other words, an explicit *techne*. This short-term "ME-thinking" appears to have led to a substantial overdesign of the pod car, with high factors of safety, that resulted in the construction of a heavy frame and the sourcing of suspension, wheels, and tires from off-road equipment, even though the ideal version of the pod car would likely not require such heavy duty materials and construction (see Fig.7.9). While not all "ME-thinking" is necessarily aligned with explicit *techne*, the materials chosen for EPTV and the structure of the design groups suggests that short-term, realizable objectives were privileged over more abstract design and production goals.

Because labor and design tasks were largely divided within the EPTV Club, the project may have also suffered from some elements of a "design by committee." Students who were further along in their coursework, or at least more eager to work on specific parts of EPTV, were allowed to choose their working groups. For example, some groups worked on the design of the passenger seats or on coding

Figure 7.9 A complete EPTV pod car frame, with suspension and electronics
Source: Author

the navigation system, only coming together to discuss work during weekly or bi-weekly meetings. Much of the work completed in these projects focused on specific objectives, though this work was also dependent upon students' personal schedules, as they also completed homework, studied for exams, and tried to have a social life outside of school. Because these groups were divided by student interest and expertise, a collaborative, messy design process – something that can assist innovative technical projects – did not occur. Even though the manufacturing group was set up specifically to move between other groups, helping design and production where it could, most work was done in relative isolation where results were reported to the entire group after being achieved. The kind of collaborative work that focuses on the broader goals of the project, an abstract *techne*, was largely absent. Perhaps because "ME-thinking" consists of a representational body of knowledge, and endeavors to be certain of specific outcomes, explicit versions of *techne* were valued over more abstract ones.

Even within the predominantly explicit *techne*-focused, small group work of the EPTV Club, attention to the *hyle* of the immediate environment was mixed. To some degree, successful designs understand the complex environments in which they are intended to function and the power of various actants in the networks they inhabit. A successful *techne* often emerges from a strong understanding of this nonhuman *hyle*. One example of the EPTV project's inattention to *hyle* involves the materials chosen for the design. Specifically, the ATV upon which EPTV pod car was based was chosen because it could be driven in an Midwestern winter. Likewise, the batteries were chosen because they were cheap, easily available, and easy to service. However, when the larger tires and the over-built suspension met with the tall, removable battery trays, the step-in height crept up over two feet. Three issues arose as a result. First, the center of gravity was raised by the additional weight. The pod car was initially designed with a low center of gravity, because of fears it could tip over or that (being on a sometimes rowdy college campus) it would be intentionally tipped over. The high battery banks, even with the weight of the batteries, chassis, and bodyshell, worked against this important design goal. Second, the two-foot step-in height would not have been a problem if all passengers were tall and pants-wearing, however, shorter, skirt-wearing passengers would have been put into a predicament trying to step into the pod car. Finally, the initial design of the pod car included a great deal of completely transparent glass, with the assumption that riders, especially female riders, would want to be seen by outsiders so as to feel safer with unknown male riders. Latour chronicles this same fear in Aramis, including the thoughts of one administration official who suggested it would be a "rape wagon" (1996, 21). Yet, EPTV designers did not consider that because the EPTV seats were roughly four feet off of the ground, female passengers would feel on "display" while riding around campus (according to insights developed by the industrial design students). This oversight may have been because nearly all EPTV Club members were male, as were the mechanical engineering undergraduate student population, the population of the course in which the EPTV design was initially conceived, and the advisors. The population of future users, however, was likely going to be much more mixed. The EPTV project did not attend particularly well to the various *hyle* of design and its impact upon future users.

Of course, it is likely that EPTV designers simply did not know enough about their potential users, and therefore the project failed to understand whether or not the EPTV was needed. Early on in the project, an industrial design course worked with the mechanical engineering students to redesign the pod car based on data from a survey of potential users. Unfortunately, I have no evidence that any of this data was incorporated into the overall design.[6] EPTV designers never sought to understand the needs of its potential passengers, nor incorporate riders' desires into the design. In much the same way that Latour suggests that Aramis failed to incorporate transformations of its environment, EPTV can be accused of not thoroughly considering the nonhumans in its own environment (1996, 280). Materials were chosen with some attention to weather, though were over-built and limited by budget constraints. Though the EPTV was inspired from the university's master plan, which focused on a specific campus and city, few decisions were coordinated with the people in that environment. A quick example of this missed coordination: EPTV pod cars were intended to run on the campus bike lanes, thus avoiding vehicular traffic. However, no bike commuters were queried about such use and it is unlikely that many cyclists would welcome the additional traffic. For Aramis, Latour explains that Parisians were not consulted about the design. He questions whether "flesh-and-blood passengers [would] subscribe to Aramis' version of them, and settle nonchalantly into the comfortable spot that the experts have spent ten years preparing for them?" (1996, 187). It seems as though EPTV was similarly designed, with a compartmentalized design process that focused on explicit *techne* production at the expense of an abstract *techne* and less focus on the specific *hyle* of the design's intended environment, including the eventual end-users.

Interfaces

In much the same way that the EPTV project should have been designed with greater consideration of the many actants that comprised its surrounding network, the level of sensitivity put toward interfaces was also ill-considered. Interfaces provide for an immersive environment that transports users and functions as the boundary where two humans and nonhumans meet and create knowledge. The interfaces in EPTV never reached the stage where they were fully tested with a group of users, so we will never know for certain how effective they were at creating new meaning. That said, designers considered the host of interactions, motions, abilities, and contexts that users would experience with the EPTV transportation network and did attempt to create necessary interactions for these hypothetical movements. Much like the Aramis project, the EPTV project did not employ a user-centered design perspective and often only hypothesized what users would want and need. While this is problematic, it in no way bars technical projects from success.

The interfaces that were created during the development of the pod car proof-of-concept exhibited some characteristics that we might see as confusing or perceptually challenging. For example, the initial design included a joystick controller. This design feature was only intended for early models, to provide a "manual" drive override mechanism.[7] The interaction between the pod car "driver" and the joystick was largely dictated by the articulation programmed into the joystick itself, though it also asked users to understand directional capacity and movement in an unusual way (see Fig. 7.10). To move the joystick left-and-back to brake may have a "natural" correlation between bodily and vehicular movement, but to press the trigger and move the joystick right-and back to engage reverse and right-and-forward to halt reverse could have been difficult for some users to intuit. That said, I do not want to suggest that this particular interface was significantly damaging to the EPTV as a kind of persuasive transportation – it was only meant for small numbers of vehicles. A single interface is not likely going to have a deleterious effect on a project, so the complex (and perhaps counterintuitive) joystick controls may not have been that significant. Other interfaces within the EPTV pod car were more likely to present negative interactions.

One predominantly featured interface present in all pod car plans was a large, red button that protruded from the dashboard. This button functioned as an emergency stop and would shut down the pod car when pressed. The metaphor that emergency stop button served to enable was: "button press equals emergency

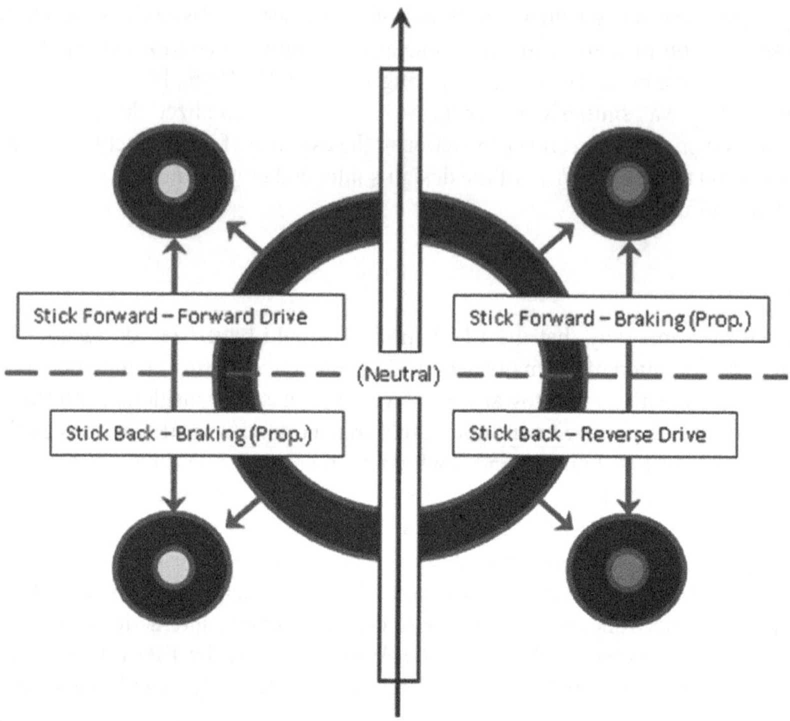

Figure 7.10 The function pattern of the pod car joystick
Source: Author and EPTV Club

stop" but also, because the button was not intended to be used very often, yet was prominently displayed, "button equals potential emergency." This metaphor traffics on the assumptions that the color red is enthymematic for danger and large red buttons are enthymematic for "all-stop." Designers determined that this button was necessary, and it was certainly a useful and potentially lifesaving interface. If an EPTV pod car were to veer out of control, the emergency stop button could bring it to a halt. However, because this interface trafficked a rather negative metaphor, passengers were likely to believe that the EPTV pod car, a new form of transportation, was unsafe. After all, most everyday forms of transportation do not prominently feature emergency controls (aside from some light rail). The result could have been a *bodily preparedness* to press the emergency stop button and the assumption that this new method of transportation was unsafe. Users would have been aware of the interface, and could have been more attentive to its location and their respective relationship to the button.

A third and final interface worth considering is that which was designed to attract the most attention from eventual users, the touch screen GUI inside the pod car. As users entered the vehicle, they were to confirm the location of their intended destination by engaging with the touch screen. However, no version of the GUI was ever paired with the coding that was to have supported it. This may have been due the shift in manufacturing focus on the pod car and the general assumption that a proof-of-concept would not need to have a fully functioning GUI. At one point, I worked with students in a technical communication course to create a web-based mock-up of the GUI. We generated and usability tested several versions of a clickable, navigable interface for the EPTV, demonstrating some design options for a touch screen interface. While the EPTV Club incorporated these designs into their plans, the project ended before they were realized. The issue with the interfaces designed by the EPTV project, then, was not that they would necessarily confuse potential users, but that they were not generated by starting with successful interface design practices and merging new interfaces with more comfortable ones. We cannot accuse the EPTV project of developing poor interfaces, but we can identify some potentially negative relations that, if tested, could prove difficult for users to engage.

Logistics

When we consider whether the EPTV project would have produced a compelling logistic, we once again run into the problem where EPTV never reached a stage of completion where it could be tested with real users. Even so, we are able to make some educated hypotheses about the assumptions built into the EPTV project and how that logistic may have played out. A logistic is how we are delivered and how we deliver ourselves, but also it is how we circulate, in repetitive ways, in our environments. Any successful logistic will take into consideration the wide array of actants that it gathers together, and if we consider human users, how they generate knowledge with interfaces. The EPTV project, however, was never really grounded to its surrounding *hyle*, even though it originated from an engineering question that was based on a university master plan. That is, the EPTV project was intended to be as autonomous as possible, with no physical connections to the ground (no rail tracks) and highly mutable in terms of physical location. Such a high degree of autonomy meant that EPTV was not tied to the physical location and therefore not particularly attendant to the immediate context in which it was intended to function. We can connect such autonomy with a lack of *hyle*-oriented thinking, but we can also identify complications in the planned logistic itself.

A successful transportation logistic should not only take the physical environment and the many actors in that environment into consideration, it should also consider the human faculty for calculation and order. When logistics are developed so people can easily engage them and when trajectories among different interfaces can be envisioned and then enacted upon, they are more likely to encourage a sustained movement. While the EPTV project never reached the phase where users were able to experience it, users would have been asked to co-ordinate the following interfaces:

1 A mental (and literal) map of where the EPTV network offered transportation
2 A sign by the EPTV pod car track (also the bike lane)

3 Users' personal cell phones to text the EPTV network the number of the stop (in time, this system was to be replaced by an app)
4 An EPTV pod car (including any and all attendant visuals, movements, and sounds)
5 Users' university ID card
6 A card reader
7 The door of the EPTV pod car
8 The touch-screen interface (see Fig. 7.11)

As a logistic, EPTV was likely to be somewhat confusing, in part because it asked users to engage with a high number of interfaces, considering the relatively short distances that EPTV was designed to cover. In order to make sense of the network, users would need a mental map of the locations on the bike path where an EPTV would appear when texted. On the surface, this does not seem to be too taxing of a proposition. Over time, this logistic could have become second nature to users if the interfaces functioned seamlessly. While we will never know for certain, I suspect that EPTV would have faced a rough path towards initial acceptance, in part because it asked users to negotiate several new interfaces in order to travel distances often less than two miles.

Figure 7.11 The touch-screen model considered for the pod car
Source: Author

EPTV is likewise difficult to evaluate, because it never reached the stage where all of the interfaces were complete, much less the stage where the interfaces could be coordinated by users. The logistic was very much still a fiction, though it has to be said, a fiction that included neither a strong commitment to understanding the immediate context of the project nor the creation and production of all necessary components. For EPTV to function as a compelling series of connected interfaces, so much depended upon the software that would connect pod cars with users, though the development of this technology was postponed until later in the design process. The focus of the course and the EPTV Club was the manufacture of the pod car, which became the central concern of weekly manufacturing and development. Perhaps a lack of attention to coding work can also be due to the "ME-thinking" displayed in the penchant for explicit *techne* activities. A stronger commitment to designing a software network could have helped the EPTV, though users would have needed to experience a number of interfaces together in to evaluate the network. While we cannot say for certain that users would have found EPTV's logistic too confusing, it did ask users to "gather" a significant number of interfaces to travel a relatively short distance.

Navigation

Making sense of the EPTV project from the perspective of the third form of *techne* explained earlier, a user *techne*, is similarly hypothetical, because users never employed EPTV as a means of transportation. Like any failed project, and much like Aramis, an honest appraisal of how EPTV was to function in its intended environment will always remain speculative. However, we are able to make some evaluative judgments about whether it would have been successful. One of the EPTV project's larger missteps was the lack of focused attention to eventual end-users. Industrial design students performed surveys, and there was some attention by the Marketing working group toward the public perception of EPTV, but not enough questions were asked of eventual end-users. Because the EPTV project was developed as a response to a planning document, and not to user need or larger social issues, the project seemed to operate on a similar level of abstraction. Of course, we can also craft a response to such criticism – EPTV was never tested with users, because the project focused so much effort on the manufacture and production of something that would *eventually* be tested with users. A physical pod car was central to generating any user feedback. Yet, users should have been part of the project from the very outset, as key aspects of the design process. Building something *and then engaging* with users is neither user-centric nor iterative. Instead of putting the cart before the horse, the EPTV project put the pod car before any user testing. Because users were only hypothetical in most of the EPTV project, their art of navigation was ignored. Users were *designed for*, such as when the seats were designed to fit 99% of body types, but not *designed with*, as when women were ignored in the design.

Nonetheless, the EPTV project made a number of assumptions about the roles that human users would play while engaging with the EPTV network. If we consider how the project conceptualized a person who wanted to travel across campus, some interesting movements are brought to the fore. Specifically, riders would have been asked to engage in relatively passive forms of bodily knowledge, or *metis*. Unlike an automobile driver, EPTV participants would not have been particularly active participants in the movement of the pod car. Aside from choosing to use the EPTV, and engaging with their cell phones, ID cards, and the touch screen, users take little action toward moving the pod car itself (the abovementioned joystick was not intended to be in all pod cars). Likewise, EPTV would have asked for propitious timing in terms of users' cell phones but little else – after all, EPTV was envisioned as an on-demand service. Very little of EPTV-based travel was intended to left to chance, *tyche*, as multiple pod cars would ensure that transportation would always be available, and its largely autonomous vehicular movement suggested that the nonhumans in the EPTV network would not need to consider, evaluate, and then respond to the chance movement of other actors. The pod car itself would be doing this of-the-moment calculation. Through a range of cameras and radar systems, EPTV was designed to avoid walkers, cyclists, and foreign objects in its intended path. Users were going to be put into a unique situation with regard to their mobility. We cannot easily judge the effectiveness of this system, but it would have asked riders to engage some unique navigational skills.

Operating under the assumption that automobiles would be less welcome on campus, the primary goal of the EPTV project was to make transportation easier for humans by ascribing greater agency to nonhumans. As such, if we were to look for a "bit" in terms of the EPTV project, we might be confused. Certainly, there is no exact "bit" in the EPTV, because a "bit" is a direct, yet simplified technological relationship between the humans and the nonhumans in a *techne* of navigation. What we might call a "bit" is shared among a number of different actants involved in the network. However, unlike an automobile, EPTV users' sense of control is less identifiable, because the "bit" is distributed away from the user and toward the systems controlled by the guidance network and pod car. Autonomy and self-guidance are not negative frames of agency; users could easily become accustomed to the more passive positions they are asked to take when riding in a pod car. Whether eventual users would be comfortable with the level of nonhuman agency in the EPTV transportation network, and whether users would experience a satisfying sense of a "bit" through their cell phones (and the eventual app), is unknown. Unfortunately, EPTV never developed to the stage where these questions could be answered.

EPTV as persuasive transportation

The result of all of these levels of inquiry, then, should tell us something about why EPTV failed; it should help us understand whether EPTV was a compelling, persuasive project, as outlined by the previous chapters. First, from the perspective of an actor–network inquiry of large transportation projects, we might consider whether EPTV suffered from a lack of identifiable phases or whether it absorbed the contradictions in definition that emerged as the project evolved.

Certainly, EPTV appeared to transition from a course project into a club project without significant issues, and while the over-reliance on critical design reviews may have led to some continuity concerns, the project did have momentum throughout. EPTV enrolled a significant number of actors, including many that cared about the project and who were involved in the technical support and production of the pod car. Certainly, the actor–network diagrams support this. Of course, the actor–network diagrams also reveal a reliance upon mechanical engineering explicit *techne* and perhaps not enough of a focus on abstract *techne* design practices or the eventual user *techne*. Such a focus is not damning, as there was likely ample time to begin user-testing once the pod car and navigation guidance systems were completed as a proof-of-concept. It is also possible that the large number of actors in the EPTV project developed competing definitions of what EPTV was supposed to be. If, like Aramis, EPTV was unable to cohere these different definitions, it may have broken apart at the seams. However, we simply do not have enough evidence to suggest that the competing definitions, of which there are many, caused a project failure.

Further, the EPTV design process was also built upon some problematic assumptions. For one, compartmentalized learning from the engineering course certainly led to isolated, specialized work within the club. EPTV may have slipped into a design-by-committee structure that may have been challenging for an innovative project. There was also a lack of attention to the specific *hyle* involved in the project and the material environment in which EPTV was to eventually function. In terms of the materials used in production, EPTV designers chose them with a recognition that weather would impact the vehicle, but not an awareness that the materials would exclude many users from comfortably using the pod cars. Some materials were chosen because they were available, with the assumption that a proof-of-concept need not represent the finished project. This is a fair assumption, though EPTV's proof-of-concept was further away from a functional prototype than it could (and possibly should) have been. Because the pod car was the public representation of EPTV (until the emergence of the remote control car), the proof-of-concept impacted public understanding. The project also made a number of assumptions about how users would acclimate to new interfaces and a new form of logistics. While these assumptions may have been perfectly acceptable, without usability testing, this uncertainty remains unarticulated. The EPTV project operated from *assumptions* about what kinds of *techne* of navigation would be compelling for users, without much consultation of whether or not users might find such a form of use to be sufficiently engaging. EPTV's construction of a largely passive form of user *techne* was speculative, and while it may have been compelling, as suggested by other autonomous projects, the lack of user consultation was troublesome, at the very least.

Latour suggested that Aramis did not succeed because no one loved the contradictions of the project nor the research necessary for the project to succeed. I believe we can state that EPTV failed because it was not a persuasive enough transportation project, from the conception of a kinesthetic rhetoric that I have outlined here. It was less attentive toward its respective actors, human and nonhuman, and focused too much on compartmentalized design methods. While the interfaces and potential logistics of the EPTV project could have been problematic, there was too great a gulf between the explicit *techne* designs generated by the students and the larger abstract *techne* needed for an innovative project. While the club's advisors may have appreciated this level of *techne*, this focus was not shared by those doing the CAD design work and building the EPTV pod car. As such, the project became aligned with a kind of "ME-thinking" about design and manufacture. Additionally, users' *techne* of navigation were less understood and not integrated with production. While it may be easier to condemn a project that is already dead, according to the concepts of a kinesthetic rhetoric, EPTV was unlikely to succeed as a persuasive transportation design. While projects similar to EPTV are likely to succeed in the near future, it failed, not because of a lack of love, but because those involved in the project failed to understand how everyday transportation can be persuasive.

EPTV and technical communication

My own involvement with EPTV began in the fall of 2010, as I was told about the project by one of my technical communication students, an enthusiastic club member. I attended a few weekly meetings and then approached the head advisor and the executive board. At the time, as a PhD student new to technical communication, I was uncertain what role I might play in the EPTV Club, though advisors and club members felt they needed help with two main communication issues: educating the public and managing documentation. In the coming year, we focused on a few activities that would make EPTV more widely known: a new logo and slogan, a series of promotional posters, more club callouts, increased numbers of public events, and continued integration with existing courses. In terms of documentation management, we began using a version control system (VCS), collected and scanned documents that existed only in hard copy, streamlined the shared network drive where materials were held, and photographed the notes that working groups would write on the workshop's blackboards. However, these strategies represent a fairly limiting conception of what technical communication scholars can bring to an innovative technical project like EPTV.

As symbolic-material workers, technical communicators can assist in articulating and facilitating many different relationships within complex transportation projects. Focusing on the aspects of a kinesthetic rhetoric, the EPTV project could have integrated technical communicators earlier on in order to facilitate the persuasive aspects of transportation. If engaged from the outset, I believe they could have assisted EPTV in a number of ways:

1 One of the largest causes for concern in the EPTV project was its lack of coordination with real users and its reliance upon largely hypothetical future users. Technical communicators could advocate for the integration of actual users, through a range of usability and user experience methods, throughout the entirety of the conceptualization and design process. A user-centered

design, one where the transportation technologies are conceived in response to direct experience and iterative feedback, would help ensure that the innovative technologies do not stray from the needs of the users.

2. Beyond the needs, desires, and safety of the human users, technical communicators could consider the roles of nonhuman materials in design. Further, the consideration of *hyle* would continuously question the relationship between the materials chosen for design and future relationships with other humans and nonhumans. Considering the impact of *hyle* in the network of actants would assist in generating both short- and long-term views of network sustainability.
3. Technical communicators could help guide the design of user interfaces, seeking out pleasing metaphors that assist comprehension, without the expectation that any single metaphor will remain static. This process would blend "dead" metaphors with novel and engaging ones that help users understand how a specific technology works. Here, the focus could involve the touch screen GUI interface, though would not be limited to that interface. The unique and attention-grabbing metaphors and interfaces would aim to promote the larger environmental goals of transportation projects.
4. As part of a broader focus on interfaces, technical communicators could work with users to understand how they make sense of the transportation system, and then bring that knowledge to bear on the design of interfaces. The concern would be on identifying how humans understand how the transportation functions, with the end goal of using that knowledge to facilitate technical mediations between humans and nonhumans. This is certainly a more difficult goal, but one that could guide production decisions.
5. Technical communicators could work alongside designers and subject matter experts throughout the manufacturing process, recording decisions and asking questions about the design choices as they are being made, with the larger goal of influencing the "scripts" that are built into the technologies. Key to this process, as with a focus on users and *hyle*, would be to bring users into the conversation from the outset. Technical communicators would not be working to "de-script" technologies after they have been developed, but shaping the kinds of expectations and affordances that they actively encourage.
6. Though we can only speculate about whether EPTV's planned logistic would have been persuasive, the number of interfaces that users were asked to negotiate was a concern. Bringing users' voices into the process, technical communicators could determine problems users might have with coordinating different interfaces. Though this work may not always be precise, it would be able to identify any large concerns before they were designed into a logistic.
7. Technical communicators could identify moments of *metis* and *kairos* in a typical act of using a transportation technology. This process would involve actual users and usability testing methods to gain feedback on the *techne* of navigation that would comprise most experiences of the transportation network. Certainly, some of these methods would seek out data that would allow technical communicators to identify where moments of *kairos* occur and how

metis is employed. The goal of this analysis would be to create conceptual models of the pod car's driver/passenger–vehicle environment.

8 Finally, working to understand agency within the act of navigation could be a larger technical communication goal. Here, researchers would hope to investigate the impact of autonomous features on human passengers and whether that sense of agency is persuasive for users. This might involve the identification of the "bit," though could also seek out conceptions of agency in general.

Some of these goals do not have specific, determinate outcomes, as seeking out a rich description of *kinesis* movement or articulation users' experience of agency is likely going to be an abstract process. However, each of these elements of a kinesthetic rhetoric promotes more satisfying and coherent forms of mobility for the major actants involved. Further, these persuasive features are even more key for fragile, innovative transportation projects that can ill afford to create technologies that fail to meet the material contexts in which they are intended for use.

In this chapter, I have attempted to bring Chapters 2–6 into greater coherence by applying the theories and methodologies from those chapters to one specific case study – the development of an all-electric autonomous pod car network at a major Midwestern university. After exploring other actor–network theory approaches to failed transportation projects, including Bruno Latour's work on Aramis, I used actor–network methods to analyze the EPTV project. While Latour suggested that Aramis failed because of a lack of love, I focused on studying EPTV from the perspective of a kinesthetic rhetoric. While EPTV was not far enough along to make certain conclusions about how it was able to satisfy its users, we can speculate that EPTV suffered from a lack of persuasive kinesthetic rhetoric. Technical communicators, working with a sense of kinesthetic rhetoric in their work, could have brought about a more successful result.

Notes

1 Initially, this path was intended to double as the existing bike paths on campus.
2 The EPTV project was also documented on at least three different Websites, none of which are still around.
3 Aside from the Challenger disaster, technical communication researchers have studied telecommunication failures, the Deepwater Horizon catastrophe, and the Ford Pinto's problematic gas tanks.
4 I conducted IRB-approved interviews with former EPTV Club members in the spring of 2010.
5 Few of the most meaningful technological projects were understood as feasible as they were developed. Though I cannot help but think of the World Wide Web as an example, Latour offers another in *Aramis* with his interview with M. Henne, head of the bureau of technological studies of *Aeroport de Paris*. Henne states, "I still tell myself that if somebody came up with the idea of the automobile today and had to go before a safety commission and explain, I don't know, let's say, how to get started on a hill . . .! Just think how complicated it is: shifting gears, using the hand brake, and so on. He wouldn't stand a chance! He'd be told: 'It can't be done.' Well, everybody knows how to start on a hill! It's the same with Aramis" (46–47).

6 I have never seen the results of this survey, nor did any members of the EPTV Club executive board. They are effectively part of the "lore" behind the early design of the EPTV. A rumor about the exterior design of the EPTV, that an advisor's college-age daughter suggested it should resemble an iPod, is also possibly apocryphal.

7 The EPTV project initially conceived of a "non-autonomous" model controlled through the joystick. I suspect that this joystick was intended as a remediating metaphor for the uncertainty that comes with a radically new interface. Transportation history is littered with these metaphors, such as early twentieth-century vehicles that included buggy whip holders.

Conclusion

This book has been my attempt to articulate a material rhetoric for persuasive transportation in order to show that persuasion plays a deep and significant role in transportation design and use. In doing so, I focused on a set of rhetorical features that play out in the creation of innovative technical projects and the everyday, habitual engagement with the end result of those same projects. I named the guiding concept for this study a "kinesthetic rhetoric" and connected persuasion with material movement as it happens through a range of human and nonhuman influences. This persuasion happens unintentionally, and most powerfully, when we are not aware we are being persuaded. Further, we can see these elements of persuasion through design processes, interface development, logistics planning, and performed acts of navigation. Because of their disciplinary history and current research interests, I suggested that technical communicators are well-positioned to function as "symbolic-material" workers in these endeavors, able to contribute to transportation research and design.

My overall goal in identifying persuasion in such a way, and in speculating on the role of technical communication researchers and practitioners in material persuasion, has been to identify new sites where technical communication can help tackle a particularly troublesome problem. Our internal-combustion automotive technologies are unlikely to survive this century, because they lead to increased greenhouse gas emissions and traffic congestion and are entirely dependent upon a dwindling nonrenewable resource. This investigation seeks to assist new and innovative transportation designs by articulating the complex interactions between humans and nonhumans, and identify how these relationships need to be sustainably persuasive beyond the short-term desire to profit off of a new technology. This environmental goal places a burden on transportation designers, who should be able to rely on technical communicators as experts in the day-to-day documentation management and as advocates for more sustainably persuasive forms of mobility. In a realm as speculative as that of innovative transportation, positioning technical communicators as symbolic-material workers is also admittedly speculative. However, when the stakes for future mobility are as high as they are, such speculation seems necessary.

In part, this book has also served as a continuation of concepts developed by Byron Hawk in a 2004 special issue of *Technical Communication Quarterly* on the value and importance of *techne*. I introduced Hawk's work in Chapter 2, as a way to ground the more theoretical discussions of unintentional, everyday, nonhuman persuasion. While Hawk's context is the technical communication classroom and the pedagogies enacted therein, his notion of a post-*techne* is more attuned to context-dependent concerns in ecologies of production. Hawk argues that *techne* "is both a rational, conscious capacity to produce and an intuitive, unconscious ability to make," essentially making the case for the value of explicit *techne* and abstract *techne* (372). Focusing on *techne* as making arts, he suggests that teachers should attend to abstract *techne* as well as the techniques of an explicit *techne*. A post-*techne*, Hawk claims, "should set up constellations of relations that allow its users [students] to see something as something else – that is, to see in a new way through those constellations of relations" (379). My goal here has been to show that mobility technologies are similarly something that can be seen by transportation design teams as *something else*, hopefully allowing us to recognize that in the movement of everyday life, new relations that can be developed, cultivated, and enacted. To engage these relations in innovative transportation projects, technical communicators should not be passive actants, but advocates for persuasive, sustainable use.

This study considers transportation to be a mundane, habitual affair – one that we are more than happy to keep habitual and mundane. In part, many in the West have learned to think of automobility as something that will always be there. The majority of the US population has access to at least one automobile and considers it an essential part of everyday life. Many other cultures and nations are developing these same relationships. In order to shape these everyday habits, the persuasive techniques that we employ also need to meet these habits where they live – at the level of unconscious motivation. If a vehicle is designed with a persuasive interface to help attain better mileage, or if a logistic for a car-sharing service persuades users to alter their transportation habits, then we are all better off for those small acts of persuasion. Perhaps we can think of these designed-in, material forms of persuasion as increasingly relevant and increasingly effective when more overt forms of persuasion fail.

Not too be overly dramatic about our current situation, but it would be more than fair to say that humans are not making enough changes to deal with the current depletion of fossil fuels or mitigate the causes of global warming. We are at the end of a roughly 200-year era of energy use called the fossil fuel era, whether we like it or not. Natural gas and oil will run out sometime in the second half of the twenty-first century and coal will follow soon after. Not only will we be forced to make massive changes in how we generate energy for heating, cooling, and producing electricity, but our methods for transportation design will be forced to change as well. The era of internal-combustion engine automobility is nearing a close and the sooner we are able to engage our efforts into practical, persuasive solutions that mitigate or alter the relationships between transportation and greenhouse gas emissions, the better. If we continue to ignore the inevitability of the changing energy landscape, or simply ignore that our energy use contributes to global warming, we play a rather dangerous game with the future of our planet's livability. Elon Musk, the outspoken CEO of Tesla, has argued that the continuance of our dependence upon fossil fuels is:

> the dumbest experiment in history. [. . .] The greater the change to the chemical composition of the physical, chemical makeup of the oceans and atmosphere [due to increased carbon emissions], the greater the long-term effect will be. Given that at some point they'll run out anyway, why run this crazy experiment to see how bad it'll be? We know it's at least some bad, and the overwhelming scientific consensus is that it'll be really bad.
>
> (Urban 2015)

How the transportation industry is able to respond to this continuing and emergent problem is likely to take one of several forms.

Adam Jonas, Managing Director at Morgan Stanley and automotive researcher, has previously made a compelling case that the effect of the sharing economy and the future of transportation will result in new systems of transportation. Defining this shift as "The Four Stages of Mobility," Jonas explains how we are still in an over 100-year long phase of mobility that features internal combustion engines, human drivers, and individually owned vehicles. The next step will take two directions: one allows the sharing economy to lead and features human drivers and low-tech car sharing services, such as Lyft and Uber. Their advantage lies in fewer emissions and less traffic, because vehicles are used only as-needed. Here, transportation technology is altered by innovations in logistics. The second option involves nonhuman drivers of individually owned vehicles, such as the autonomous cars that Apple, Google, Tesla, and scores of other companies are rapidly developing. These would involve similar savings in traffic and emissions – even more so if these vehicles are propelled through alternative fuel sources. The final stage, though, the one that Jonas sees as inevitable, is a combination of these two – autonomous, shared vehicles that will likely feature alternative fuel delivery systems (Morgan Stanley 2015). These transportation networks, currently in a deeply experimental phase, would require designers to call upon both concrete and abstract design methods, engage users in reassuring interfaces, require compelling logistics, and offer repeatable forms of shared agency in navigation. These concerns are likewise ones that the EPTV project either faced or would have encountered during its development and ones that every vehicle must encounter on its way through production and to eventual use. We are at a moment in history when we must question the once standardized methods for vehicle design and pay attention to what Hawk calls the "constellations of relations" that allow us to "see something as something else" – to see vehicles for what they once were and could eventually become (379).

Becoming a technical communicator within a rapidly changing technological landscape and amidst all of these uncertain environmental issues is not an easy task. Those who wish to take responsibility for doing this kind of work in innovative transportation projects, like the ones described here, will need to acknowledge and emphasize the inherent material persuasions that they are a part of and can help articulate. These symbolic-material workers will be doing more than developing strategies for content management, documentation, memo writing, or meeting-arranging, but instead hoping to take part in a holistic design process. That said, we need other workers in design teams to understand that technical communicators' roles can eclipse the transcription or translation of technical material, but also include the articulation of how symbolic designs meet material realities. Further, I believe these concepts can be conveyed to students in technical communication programs, as they enter the job market as writers and communication strategists, certainly, but also as professionals ready to be key personnel in innovative design projects. We might align these roles to that of an "experience architect" (see Potts 2013; Salvo 2014), and identify as technologists, social scientists, and rhetoricians deeply concerned with the sustainability of transportation and future of human mobility. Potts argues that experience architects "can lead the analysis and design for both the social and technological infrastructures of communication by framing new practices, creating new tools, and enabling richer and more valuable interactions" (2013, 4). The variety and impact of technical communication can greatly expand in the coming years, and take on such leadership roles, but only if we acknowledge that these new challenges are within our expertise and ability.

The work of technical communication in this context is challenging, to say the least, as there are numerous factors far beyond our control. However, if we want to envision change in a system so deeply ingrained into everyday life, we need to embrace new methods and new positions that understand and engage the relations between materiality, movement, and persuasion. Internal combustion engine automobility is a wicked problem in need of a solution and technical communicators can be part of the solution, working to change accepted ways of thinking and articulate the role of material persuasion transportation design.

References

Ackrill, J. L. 1997. *Essays on Plato and Aristotle*. Oxford: Oxford University Press.

Akrich, Madeline. 1992. "The De-Scription of Technical Objects." In *Shaping Technology/Building Society: Studies in Sociotechnical Change*, edited by Wiebe E. Bijker and John Law, 205–224. Cambridge, MA: MIT Press.

Albers, M. J. 2003. "Single Sourcing and the Technical Communication Career Path." *Technical Communication* 50.3: 335–344.

Alexander, G. J. and H. Lunenfeld. 1986. "Driver Expectancy in Highway Design and Traffic Operations." In *US Department of Transportation Report No: FHWA-TO-86–1*. http://ntl.bts.gov/lib/32000/32500/32510/FHWA-TO-86–1.

Amin, Ash and Nigel Thrift. 2002. *Reimagining the Urban*. Cambridge: Polity Press.

Anderson, Rebekka. 2014. "Rhetorical Work in the Age of Content Management: Implications for the Field of Technical Communication." *Journal of Business and Technical Communication* 28.2: 115–157.

Aristotle. 1984. In *The Complete Works of Aristotle*, edited by Jonathan Barnes. Oxford: Oxford University Press.

Arrian of Nicomedia. 2013. *Alexander the Great: The Anabasis and the Indica*. Translated by Martin Hammond. Oxford: Oxford University Press.

Atwill, Janet M. 1998. *Rhetoric Reclaimed: Aristotle and the Liberal Arts Tradition*. Ithaca, NY: Cornell University Press.

———. 2006. "Bodies and Art." *Rhetoric Society Quarterly* 36.1: 165–170.

Barthes, Roland. 1957/1972. *Mythologies*. Translated by Annette Lavers. New York: Hill & Wang.

Bazerman, Charles. 1994. "Systems of Genre and the Enactment of Social Intentions." In *Genre and the New Rhetoric*, edited by A. Freedman and P. Medway, 79–99. London: Taylor & Francis.

Biesecker, Barbara A. and John Louis Lucaites. 2009. "Introduction." In *Rhetoric, Materiality, and Politics*, edited by Barbara A. Biesecker and John Louis Lucaites, 1–16. New York: Peter Lang Publishing.

Bilger, Burkhard. 2013. "Auto-Correct: Has the Self-Driving Car at Last Arrived?" *New Yorker*, November 25. www.newyorker.com/magazine/2013/11/25/auto-correct.

Billmaier, James. 2010. *JOLT!: The Impending Dominance of the Electric Car and Why America Must Take Charge*. Charleston, SC: Advantage.

Birch, Douglas and John H. Fielder. 1994. *The Ford Pinto Case: A Study in Applied Ethics, Business, and Technology*. Albany, NY: State University of New York Press.

Bitzer, Lloyd F. 1968. "The Rhetorical Situation." *Philosophy and Rhetoric* 1.1: 1–14.

Blackwell, Alan F. 2006. "The Reification of Metaphor as a Design Tool." *ACM Transactions on Computer-Human Interaction* 13.4: 490–530.

Blake, Eric S., Todd B. Kimberlain, Robert J. Berg, John P. Cangialosi, and John L. Beven II. 2013. "Tropical Cyclone Report, Hurricane Sandy." *National Hurricane Center*, February, 12. www.nhc.noaa.gov/data/tcr/AL182012_Sandy.pdf.

Böhm, Steffen, Campbell Jones, Chris Land, and Matt Peterson. 2006. "Impossibilities of Automobility." In *Against Automobility*, edited by Steffen Böhm, Campbell Jones, Chris Land, and Matthew Paterson, 3–16. Oxford: Wiley-Blackwell.

Broder, John M. 2013a. "Stalled Out on Tesla's Electric Highway." *New York Times*, February 28. www.nytimes.com/2013/02/10/automobiles/stalled-on-the-ev-highway.html?ref=automobiles&_r=1&.

———. 2013b. "That Tesla Data: What It Says and What It Doesn't." *New York Times*, February 14. http://wheels.blogs.nytimes.com/2013/02/14/that-tesla-data-what-it-says-and-what-it-doesnt/.

Brooke, Collin Gifford. 2009. *Lingua Fracta: Toward a Rhetoric of New Media*. Cresskill, NJ: Hampton Press.

Bryant, Levi R. 2011. *The Democracy of Objects*. Ann Arbor, MI: Open Humanities Press.

Buchanan, Richard. 1995. "Rhetoric, Humanism, and Design." In *Discovering Design: Explorations in Design Studies*, edited by Richard Buchanan and Victor Margolin, 23–68. Chicago: University of Chicago Press.

———. 2001. "Design and the New Rhetoric." *Philosophy and Rhetoric* 34.3: 183–206.

Butler, Judith. 1993. *Bodies That Matter: On the Discursive Limits of Sex*. New York: Routledge.

Callon, Michel. 1986. "The Sociology of an Actor-Network: The Case of the Electric Vehicle." In *Mapping the Dynamics of Science and Technology*, edited by Michel Callon, John Law, and Arie Rip, 19–34. London: Macmillan Press.

Carnegie, Teena A. M. 2009. "Interface as Exordium: The Rhetoric of Interactivity." *Computers and Composition* 26.3: 164–173.

Carroll, J. M. and R. L. Mack. 1985. "Metaphor, Computing Systems, and Active Learning." *International Journal of Man-Machine Studies* 22.1: 39–57.

Carroll, J. M., R. L. Mack, and W. A. Kellogg. 1988. "Interface Metaphors and User Interface Design." In *Handbook of Human-Computer Interaction*, edited by M. Helander, 67–85. Amsterdam: Elsevier Science Publishers.

Carroll, J. M. and M. B. Rosson. 1994. "Putting Metaphors to Work." In *Graphics Interface '94, Proceedings of the 1994 Graphics Interface Conference*, edited by Wayne A. Davis and Barry Joe, 112–119. Toronto: Canadian Information Processing Society.

Carter, Joyce Locke. 2003. "The Implications of Single Sourcing on Writers and Writing." *Technical Communication* 50.3: 1–4.

Chevrolet. 2010. "2011 Chevy Volt – 'How to' Video – Dashboard | Instrument Cluster Display | Chevrolet." *YouTube*, September 28. www.youtube.com/watch?v=xtMACEqCgDg.

Childs, David. 2013. "Doctor Werner Lang: Engineer Who Designed the Trabant Car." *The Independent*, July 7. www.independent.co.uk/news/obituaries/doctor-werner-lang-engineer-who-designed-the-trabant-car-8693630.html.

Cicero, Marcus Tullius. 2001. *Cicero: On the Ideal Orator*. Oxford: Oxford University Press.

Clark, David P. 2008. "Content Management and the Separation of Presentation and Content." *Technical Communication Quarterly* 17.1: 35–60.

Clough, Patricia Ticineto. 2000. *Autoaffection: Unconscious Thought in the Age of Teletechnology*. Minneapolis, MN: University of Minnesota.

ComputerUser Dictionary. s.v. n.d. "Interface." Last accessed September 17, 2013, www.computeruser.com/dictionary?name_directory_startswith=I.

Connors, Robert. 1983. "Actio: A Rhetoric of Manuscripts." *Rhetoric Review* 2.1: 64–73.

Cooper, Alan. 1995. "The Myth of Metaphor." *Visual Basic Programmer's Journal* 5.6: 127–128.

Cresswell, Tim. 2006. *On the Move: Mobility in the Modern, Western World*. New York: Routledge.

Cross, Nigel. 1999. "Design Research: A Disciplined Conversation." *Design Issues* 15.2: 5–10. Print.

———. 2001. "Designerly Ways of Knowing: Design Discipline versus Design Science." *Design Issues* 17.3: 49–55.

———. 2006. *Designerly Ways of Knowing*. London: Springer-Verlag.

Dant, Tim. 2004. "The Driver-Car." *Theory, Culture & Society* 21.4/5: 61–79.

DARPA. n.d. "Grand Challenge Overview." Last accessed August 13, 2014. http://archive.darpa.mil/grandchallenge04/overview.htm.

Davis, Diane. 2010. *Inessential Solidarity: Rhetoric and Foreigner Relations*. Pittsburgh, PA: University of Pittsburgh Press.

Dayton, David and Keith Hopper. 2010. "Single Sourcing and Content Management: A Survey of STC Members." *Technical Communication* 57.4: 375–397.

DeCicco, John and Freda Fung. 2006. "Global Warming on the Road: The Climate Impact of America's Automobiles." *Environmental Defense Fund*, April 22, 2014. www.edf.org/sites/default/files/5301_Globalwarmingontheroad_0.pdf.

Dennis, Kingsley and John Urry. 2009a. *After the Car*. Cambridge: Polity.

———. 2009b. "Post-Car Mobilities." In *Car Troubles: Critical Studies of Automobility and Auto-Mobility*, edited by Jim Conley and Arlene Tigar McLaren, 235–252. Farnham, England: Ashgate Publishing.

Dery, Mark. 2006. "Always Crashing the Same Car: A Head-On Collision with the Technosphere." In *Against Automobility*, edited by Steffen Böhm, Campbell Jones, Chris Land, and Matthew Paterson, 223–239. Oxford: Wiley-Blackwell.

Detienne, Marcel and Jean-Pierre Vernant. 1974/1978/1991. *Cunning Intelligence in Greek Culture and Society*. Translated by Janet Lloyd. Chicago, IL: University of Chicago Press.

DeVoss, Dànielle Nicole and James E. Porter. 2006. "Why Napster Matters to Writing: File-sharing as a New Ethic of Digital Delivery." *Computers and Composition* 23.2: 178–210.

Dragga, Sam. 1992. "The Ethics of Delivery." In *Rhetorical Memory and Delivery: Classical Concepts for Contemporary Composition and Communication*, edited by John Frederick Reynolds, 79–95. Hillsdale, NJ: Lawrence Erlbaum.

Dunne, Joseph. 1993. *Back to the Rough Ground: 'Phronesis' and 'Techne' in Modern Philosophy and in Aristotle*. Notre Dame, IN: University of Notre Dame Press.

Edbauer, Jennifer. 2005. "Unframing Models of Public Distribution: From Rhetorical Situation to Rhetorical Ecologies." *Rhetoric Society Quarterly* 35.4: 5–24.

Edmunds, Lowell. 1975. *Chance and Intelligence in Thucydides*. Cambridge, MA: Harvard University Press.

Eisenstein, Paul. 2009. "2010 Ford Fusion Sport has Crisp Design as Hybrids Get Eco-Geek Dashboard Tech: L.A. Auto Show Preview." *Popular Mechanics*, October 1. www.popularmechanics.com/cars/news/4289684.

Enos, Theresa. ed. 1996. *Encyclopedia of Rhetoric and Composition: Communication from Ancient Times*. New York: Routledge "George Campbell and the General Audience Construct."

Evia, Carlos, Sean Healy, and Tim Lockridge. 2013. "Evaluating a Workflow for Authoring Multimodal DITA." *SIGDOC '13. Proceedings of the 31st ACM International Conference on Design of Communication,* 185–186. New York: ACM.

Feenberg, Andrew. 2004. *Heidegger and Marcuse: The Catastrophe and Redemption of History*. New York: Routledge.

Fitchard, Kevin. 2012. "Ford is Ready for the Autonomous Car: Are Drivers?" *Gigaom*, April 9. https://gigaom.com/2012/04/09/ford-is-ready-for-the-autonomous-car-are-drivers/.

Fleckenstein, Kristie, Clay Spinuzzi, Rebecca Rickly, and Carole Clarke Papper. 2008. "The Importance of Harmony: An Ecological Metaphor for Writing Research." *College Composition and Communication* 60.2: 388–419.

Fleming, David. 1998. "Rhetoric as a Course of Study." *College English* 61.2: 169–191.

Flemisch, Frank, C. Adams, S. Conway, K. Goodrich, M. Palmer, and P. Schutte. 2003. "The H-Metaphor as a Guideline for Vehicle Automation and Interaction." In *Technical Memorandum No. NASA/TM—2003–212672*. Hampton, VA: LaRC. http://ntrs.nasa.gov/archive/nasa/casi.ntrs.nasa.gov/20040031835.pdf.

Fletcher, Seth. 2011. *Bottled Lightning: Superbatteries, Electric Cars, and the New Lithium Battery.* New York: Hill & Wang.

Flusser, Vilém. 1999. *The Shape of Things: A Philosophy of Design*. London: Reaktion Books.

Ford, Henry and Eugene T. Gregorie. 1942. *Automobile Chassis Construction*. US Patent 2,269,452, filed July 27, 1940, issued January 13, 1942.

Ford Motor Company Newsroom. 2008. "Ford's SmartGauge with EcoGuide Coaches Drivers to Maximize Fuel Efficiency on New Fusion Hybrid." Last updated October 29, 2008. Last accessed September 21, 2011. http://media.ford.com/article_display.cfm?article_id=29300.

Foucault, Michel. 1972. *Archeology of Knowledge*. Translated by A. M. Sheridan Smith. New York: Pantheon Books.

Fraiberg, Steven. 2013. "Reassembling Technical Communication: A Framework for Studying Multilingual and Multimodal Practices in Global Contexts." *Technical Communication Quarterly* 22.1: 10–27.

Frost, Erin, A. 2013. "Transcultural Risk Communication on Dauphin Island: An Analysis of Ironically Located Responses to the Deepwater Horizon Disaster." *Technical Communication Quarterly* 22.1: 50–66.

Gardner, Howard. 1983. *Frames of Mind: The Theory of Multiple Intelligences*. New York: Basic Books.

Geisler, Cheryl. 2004. "How Ought We to Understand the Concept of Rhetorical Agency?: Report from the ARS." *Rhetoric Society Quarterly* 34.3: 9–17.

Geron, Tomio. 2012. "Ride Sharing Startups get California Cease and Desist Letters." *Forbes*, October 8. www.forbes.com/sites/tomiogeron/2012/10/08/ride-sharing-startups-get-california-cease-and-desist-letters/.

Gibson, James J. and Laurence E. Crooks. 1938. "A Theoretical Field-Analysis of Automobile Driving." *The American Journal of Psychology* 51.3: 453–471.

Gitelman, Lisa. 2006. *Always Already New: Media, History, and the Data of Culture*. Cambridge: MIT University Press.

Google. 2015. "Google Self-Driving Car Project Monthly Report." June. http://static.googleusercontent.com/media/www.google.com/en//selfdrivingcar/files/reports/report-0615.pdf.

Gorgias. 2003. "Encomium of Helen." In *The Greek Sophists*, translated by John Dillon and Tania Gergel, 76–84. London: Penguin Classics.

Grabill, Jeffrey T. 2003. "On Divides and Interfaces: Access, Class, and Computers." *Computers & Composition* 20.4: 455–472.

Graham, S. Scott. 2009. "Agency and the Rhetoric of Medicine: Biomedical Brain Scans and the Ontology of Fibromyalgia." *Technical Communication Quarterly* 18.4: 376–404.

Graham, Stephen. 1997. "Imagining the Real-Time City." in *Imagining Cities: Scripts, Signs, Memories*, edited by Sallie Westwood and John Williams, 31–49. New York: Routledge.

Graham, Steven and Simon Marvin. 2001. *Splintering Urbanism: Networked Infrastructures, Technological Mobilities and the Urban Condition*. New York: Routledge.

Grassi, Ernesto. 2001. *Rhetoric as Philosophy: The Humanist Tradition*. Carbondale, IL: Southern Illinois University Press.

Grescoe, Taras. 2012. *Straphanger: Saving Our Cities and Ourselves from the Automobile*. New York: Times Books.

Grigar, Dene. 2005. "Kineticism, Rhetoric, and New Media Artists." *Computers and Composition* 22: 105–112.

Grimaldi, William M. A. 1972. *Studies in the Philosophy of Aristotle's Rhetoric*. Weisbaden, Germany: Franz Steiner Verlag.

Gross, Alan G. 2008. "What Aristotle Meant by Rhetoric." In *Rereading Aristotle's Rhetoric*, edited by Alan G. Gross and Author E. Walzer, 24–37. Carbondale, IL: Southern Illinois University Press.

Gross, Bill. 2013. "Google's Self-Driving . . ." *Twitter*, April 29. https://twitter.com/Bill_Gross/status/329069954911580160.

Guzzo, Erico. 2011. "How Google's Self-Driving Car Works." *IEEE Spectrum*, October 18. http://spectrum.ieee.org/automaton/robotics/artificial-intelligence/how-google-self-driving-car-works.

Haefner, Joel. 2009. "Letter from the Guest Editor." *Computers and Composition* 26.3: 135–137.

Hailey, David. 2011. "Combining Rhetorical Theory with Usability Theory to Evaluate Quality of Writing in Web-Based Texts." In *Usability of Complex Information Systems: Evaluation of User Interaction*, edited by Michael J. Albers and Brian Still, 17–45. New York: CRC Press.

Haraway, Donna. 1991. "The Promises of Monsters: A Regenerative Politics for Inappropriated Others." In *Cultural Studies,* edited by Lawrence Grossberg, Cary Nelson, and Paula Treichler, 295–337. New York: Routledge.

Hart-Davidson, William, Grace Bernhardt, Michael McLeod, Martine Rife, and Jeffrey T. Grabill. 2007. "Coming to Content Management: Inventing Infrastructure for Organizational Knowledge Work." *Technical Communication Quarterly* 17.1: 10–34.

Hawhee, Debra. 2005. *Bodily Arts: Athletics and Rhetorics in Ancient Greece*. Austin, TX: University of Texas Press.

Hawk, Byron. 2004. "Toward a Post-Techne–or, Inventing Pedagogies for Professional Writing." *Technical Communication Quarterly* 13.4: 371–392.

Hayles, N. Katherine. 2002. *Writing Machines*. Cambridge, MA: MIT University Press.

Heidegger, Martin. 2007. *Basic Concepts in Ancient Philosophy*. Translated by Richard Rojcewicz. Bloomington, IN: Indiana University Press.

The Henry Ford. n.d. "The Model T." Last accessed June 16, 2014. www.hfmgv.org/exhibits/showroom/1908/model.t.html.

Herodotus. 2008. *The Histories*. Translated by Robin Waterfield. Oxford: Oxford University Press.

Hicks, Michael. 2013. "img_7530-raw." *Flickr*, October 31. www.flickr.com/photos/mulad/10611832484.

Hirtle, S. C. 1998. "The Cognitive Atlas: Using GIS as a Metaphor for Memory." In *Spatial and Temporal Reasoning in Geographic Information Systems*, edited by M. Egenhofer and R. Golledge, 263–271. Oxford: Oxford University Press.

Hobbs, Nicholas Kenneth and Liang-Yu Chi. 2013. *Gesture-Based Automotive Controls*. US Patent 20,130,261,871, filed April 2, 2012.

Hollnagel, E. and D. D. Woods. 2005. *Joint Cognitive Systems: Foundations of Cognitive Systems Engineering*. Boca Raton, FL: Taylor & Francis.

IDEO. 2009. "Hybrid Electric Vehicle Interaction for Ford Motor Company." Last accessed November 12, 2014. www.ideo.com/work/hybrid-electric-vehicle-dashboard-interaction/.

Ijsseling, Samuel. 2008. *Rhetoric and Philosophy in Conflict: A Historical Survey*. The Hague: Martinus Nijhoff.

"interface." n.d. *OED Online*. Oxford University Press. Last accessed December 13, 2014. www.oed.com.ezproxy.proxy.library.oregonstate.edu/view/Entry/97747?rskey=yGIfyP&result=1.

Intergovernmental Panel on Climate Change. 2007. *Climate Change 2007: Synthesis Report*. Last accessed March 15, 2010. www.ipcc.ch/publications_and_data/publications_ipcc_fourth_assessment_report_synthesis_report.htm.

———. 2014. "Chapter 8: Transport." In *Contribution of Working Group III to the Fifth Assessment Report of the Intergovernmental Panel on Climate Change, Climate Change 2014: Mitigation of Climate Change*, edited by O. Edenhofer, R. Pichs-Madruga, Y. Sokona, E. Farahani, S. Kadner, K. Seyboth, A. Adler, I. Baum, S. Brunner, P. Eickemeier, B. Kriemann, J. Savolainen, S. Schlömer, C. von Stechow, T. Zwickel, and J. C. Minx, 599–670. Cambridge: Cambridge University Press.

Jain, Juliet. 2006. "Bypassing and WAPing: Reconfiguring Timetables for 'Real-Time' Mobility." In *Mobile Technologies of the City*, edited by Mimi Sheller and John Urry, 79–101. New York: Routledge.

James, Kyle. 2007. "Go, Trabi, Go! East Germany's Darling Car turns 50." *Deutsche Welle*, May 19. www.dw.de/go-trabi-go-east-germanys-darling-car-turns-50/a-2542584-1.

Jensen, Cheryl. 2011. " 'Aggravating' MyFord Touch Sends Ford Plummeting in J.D. Power Quality Survey." *New York Times*, June 23. http://wheels.blogs.nytimes.com/2011/06/23/aggravating-myford-touch-sends-ford-plummeting-in-j-d-power-quality-survey/?_php=true&_type=blogs&_r=0.

Jensen, Ole. B. 2013. *Staging Mobilities*. New York: Routledge.

Johnson, Robert R. 1998. *User-Centered Technology: A Rhetorical Theory for Computers and Other Mundane Artifacts*. Albany: State University of New York Press.

———. 2010. "The Ubiquity Paradox: Further Thinking on the Concept of User-Centeredness." *Technical Communication Quarterly* 19.4: 335–351.

Johnson, Robert R. and Frances J. Ranney. 2009. "Recovering Techne." *Technical Communication Quarterly* 11.2: 237–239.

Johnson, Robert R., Michael J. Salvo, and Meredith W. Zoetewey. 2007. "User-Centered Technology in Participatory Culture: Two Decades 'Beyond a Narrow Conception of Usability Testing.'" *IEEE Transactions on Professional Communication* 50.4: 320–332.

Johnson-Eilola, Johndan. 2005. *Datacloud: Toward a New Theory of Online Work*. Cresskill, NJ: Hampton Press.

Karrer, Katja, Susan Briest, Thomas Vohringer-Kuhnt, Thorb Baumgarten, and Robert Schleicher. 2005. "Driving without Awareness" In *Traffic and Transport Psychology: Theory and Application*, edited by Geoffrey Underwood, 455–470. Bingly, UK: Emerald.

Keats, John. 1958. *The Insolent Chariots*. Philadelphia, PA: J.B. Lippincott & Co.

Kennedy, George A. 1980. *Classical Rhetoric and Its Christian and Secular Tradition from Ancient to Modern Times*. Chapel Hill, NC: University of North Carolina Press.
———. 1992. "A Hoot in the Dark: The Evolution of General Rhetoric." *Philosophy and Rhetoric* 25.1: 1–21.
Kenward, Alyson, Daniel Yawitz, and Urooj Raja. 2013. "Sewage Overflows from Hurricane Sandy." *Climate Central,* March 26. www.climatecentral.org/pdfs/Sewage.pdf.
Kimball, Miles. 2006. "Cars, Culture, and Tactical, Technical Communication." *Technical Communication Quarterly* 15.1: 67–86.
Kinneavy, James. 1986. "Kairos: A Neglected Concept in Classical Rhetoric." In *Rhetoric and Praxis: The Contribution of Classical Rhetoric to Practical Reasoning*, edited by Jean Dietz Moss, 79–105. Washington, DC: Catholic University Press.
Kinneavy, James and Catherine Eskin. 1994. "Kairos in Aristotle's Rhetoric." *Written Communication* 11.1: 131–142.
Kirby, John T. 1997. "Aristotle and Metaphor." *American Journal of Philology* 118.4: 517–554.
Kirsch, David A. 2000. *The Electric Vehicle and the Burden of History*. New Brunswick, NJ: Rutgers University Press.
Lakoff, George and Mark Johnson. 1980. *Metaphors We Live By*. Chicago: University of Chicago Press.
Lanham, Richard A. 1991. *A Handlist of Rhetorical Terms, 2nd ed.* Berkeley, CA: University of California Press.
Latour, Bruno. 1988. *Science in Action: How to Follow Scientists and Engineers Through Society*. Cambridge, MA: Harvard University Press.
———. 1993a. *We Have Never Been Modern*. Translated by Catherine Porter. Cambridge: Harvard University Press.
———. 1993b. "Ethnography of a High Tech Case: About Aramis." In *Technological Choices: Transformations in Material Cultures Since the Neolithic*, edited by P. Lemonnier, 372–398. London: Routledge.
———. 1994. "On Technical Mediation." *Common Knowledge* 3.2: 29–64.
———. 1996. *Aramis, or the Love of Technology*. Cambridge: Harvard University Press.
———. 1999. "A Collective of Humans and Nonhumans." In *Pandora's Hope: Essays on the Reality of Science Studies*, 174–215. Cambridge, MA: Harvard University Press.
———. 2004. *The Politics of Nature: How to Bring the Sciences into Democracy*. Translated by Catherine Porter. Cambridge, MA: Harvard University Press.
———. 2005. "From Realpolitik to Dingpolitik – or How to Make Things Public." In *Making Things Public: Atmospheres of Democracy*, edited by Bruno Latour and Peter Weibel, 14–44. Cambridge, MA: MIT Press.
———. 2007. *Reassembling the Social: An Introduction to Actor-Network-Theory*. Oxford: Oxford University Press.
———. 2008. "A Cautious Prometheus? A Few Steps Toward a Philosophy of Design (with Special Attention to Peter Sloterdijk)." In *Proceedings of the 2008 Annual International Conference of the Design History Society*, edited by Fiona Hackne, Jonathan Glynne, and Viv Minto. 1–13. www.bruno-latour.fr/sites/default/files/112-DESIGN-CORNWALL-GB.pdf.
Lauer, Janice M. 1994. "Memories of Jim Berlin." *Journal of Advanced Composition* 14.2: 583–589.
Law, John and Michel Callon. 1992. "The Life and Death of an Aircraft: A Network Analysis of Technical Change." In *Shaping Technology/Building Society: Studies in*

Sociotechnical Change, edited by Wiebe E. Bijker and John Law, 21–52. Cambridge, MA: MIT Press.

Leclerc, Ivor. 2004. *The Nature of Physical Existence*. Oxford, UK: Routledge.

Liddell, Henry George and Robert Scott. 1996. *A Greek-English Lexicon, 9th ed.*, edited by Henry Stuart Jones. Oxford: Oxford University Press.

Local Motors. 2010. "How Local Motors Works." *YouTube*, June 23. www.youtube.com/watch?v=azCRuwtE_n0.

———. 2013. "Design Studio." Last accessed January 8. www.local motors.com/studio Home.php.

Lundberg, Christian and Joshua Gunn. 2005. " 'Ouija Board, Are There Any Communications?': Agency, Ontotheology, and the Death of the Humanistic Subject, or, Continuing the ARS Conversation." *Rhetoric Society Quarterly* 35.4: 83–105.

Lutz, Catherine and Anne Lutz Fernandez. 2010. *Carjacked: The Culture of the Automobile and Its Effect on our Lives.* New York: Palgrave Macmillan.

Lynch, Kevin. 1960. *The Image of the City*. Cambridge, MA: MIT University Press.

Mackiewicz, Jo. 2010. "The Co-Construction of Credibility in Online Product Reviews." *Technical Communication Quarterly* 19.4: 403–426.

Maguire, Eleanor A., David G. Gadian, Ingrid S. Johnsrude, Catriona D. Good, John Ashburner, Richard S. J. Frackowiak, and Christopher D. Frith. 2000. "Navigation-Related Structural Change in the Hippocampi of Taxi Drivers." *Proceedings of the National Academy of Sciences of the United States of America* 98.8: n.p.

Maguire, Eleanor A., Katherine Woolett, and Hugo J. Spiers. 2006. "London Taxi Drivers and Bus Drivers: A Structural and Neuropsychological Analysis." *Hippocampus* 16: 1091–1101.

Mantle, Jonathan. 1995. *Cars Wars: Fifty Years of Greed, Treachery, and Skullduggery in the Global Marketplace*. New York: Arcade Publishing.

Marback, Richard. 2008. "Unclenching the Fist: Embodying Rhetoric and Giving Objects their Due." *Rhetoric Society Quarterly* 38.1: 46–65.

Markoff, John. 2010. "Google Cars Drive Themselves, in Traffic." *New York Times*, October 9. www.nytimes.com/2010/10/10/science/ 10google.html?pagewanted=all&_r=1&.

Marshall, Sean. 2008. "PRT 3." *Flickr*, November 21. www.flickr.com/photos/7119320@N05/3057888160.

McCarthy, Jacob E., Jeffrey T. Grabill, William Hart-Davidson, and Michael McLeod. 2011. "Content Management in the Workplace: Community, Context, and a New Way to Organize Writing." *Journal of Business and Technical Communication* 25.4: 367–395.

McConnell, Steve. 2004. *Code Complete, 2nd ed.* Redmond, WA: Microsoft Press.

McCorkle, Ben. 2012. *Rhetorical Delivery as Technological Discourse: A Cross-Historical Study*. Carbondale, IL: Southern Illinois University Press.

McGee, Michael Calvin. 1982/2009. "A Materialist's Conception of Rhetoric." In *Rhetoric, Materiality, and Politics*, edited by Barbara A. Biesecker and John Louis Lucaites, 17–42. New York: Peter Lang Publishing.

McGovern, Heather. 2005. "Remembering and Applying Non-usability Testing Methods for Learning How Web Sites Function." *Technical Communication* 52.2: 175–186.

McKeon, Richard. 1946. "Aristotle's Conception of Language and the Arts of Language." *Classical Philology* 41.4: 193–206.

McLuhan, Marshall. 1964. *Understanding Media: The Extensions of Man*. New York: McGraw Hill.

McShane, Clay. 1994. *Down the Asphalt Path: The Automobile and the American City*. New York: Columbia University Press.

Media.Ford.com. 2010. "2010 Ford Fusion SmartGauge with EcoGuide Coaches Drivers to Maximum Fuel Efficiency." Last updated March, 2010, last accessed March 15, 2011. www.media.ford.com/images/10031/SmartGauge.pdf.

Meillassoux, Quentin. 2008. *After Finitude: An Essay on the Necessity of Contingency*. Translated by Ray Brassier. London: Continuum.

Michon, J. A. 1985. "A Critical Review of Driver Behaviour Models: What Do We Know? What Should We Do?" In *Human Behaviour and Traffic Safety*, edited by L. A. Evans and R. C. Schwing, 487–525. New York: Plenum Press.

Miles, Walter. 1929. "Sleeping with the Eyes Open." *Scientific American* 140: 489–492.

Miller, Carolyn R. 1989. "What's Practical about Technical Writing?" In *Technical Writing: Theory and Practice*, edited by Bertie E. Fearing and W. Keats Sparrow, 14–24. New York: Modern Language Association Press.

———. 1992. "Kairos in the Rhetoric of Science." In *A Rhetoric of Doing: Essays on Written Discourse in Honor of James L. Kinneavy*, edited by Stephen P. Witte, Neil Nakadate, and Roger D. Cherry, 310–327. Carbondale, IL: Southern Illinois University Press.

———. 2007. "What Can Automation Tell Us about Agency?" *Rhetoric Society Quarterly* 37.2: 137–157.

Mirel, Barbara. 2003. *Interaction Designs for Complex Problem Solving: Developing Useful and Usable Software*. San Francisco, CA: Morgan Kaufmann.

Mitcham, Carl. 2001. "Dasein Versus Design: The Problematics of Turning Making into Thinking." *International Journal of Technology and Design Education* 11.1: 27–36.

Moeller, Ryan and McAllister, Ken. 2002. "Playing with Techne: A Propaedeutic for Technical Communications." *Technical Communication Quarterly* 11.2: 185–206.

Mom, Gijs. 2004. *The Electric Vehicle: Technology and Expectations in the Automobile Age*. Baltimore: Johns Hopkins University Press.

Morgan Stanley. 2015. "Autos & Auto-Related." *Morgan Stanley Research,* April 7. http://orfe.princeton.edu/~alaink/SmartDrivingCars/PDFs/MorganStanley%20040715Report Jonas.pdf.

Musk, Elon. 2013a. "NYTimes Article about Tesla . . ." *Twitter*, February 11. https://twitter.com/elonmusk/status/301049593385340928.

———. 2013b. "A Most Peculiar Test Drive." *Tesla Blog*, February 13. www.teslamotors.com/blog/most-peculiar-test-drive.

———. 2014. "All our Patent are Belong to You." *Tesla Blog*, June 12. www.teslamotors.com/blog/all-our-patent-are-belong-you.

Mutchler, Tom. 2012. "2013 Ford Flex Received Updates, but Only Some Were Improvements." *Consumer Reports*, September 20. www.consumerreports.org/cro/news/2012/09/2013-ford-flex-received-updates-but-only-some-were-improvements/index.htm.

Mutzabaugh, Ben. 2013. "Wednesday: Sandy's Flights Cancellations Now at 19,500." *USA Today*, October 31. www.usatoday.com/story/todayinthesky/2012/10/30/tuesday-sandy-delay-tally/1668431/.

Nader, Ralph. 1965. *Unsafe at Any Speed: The Designed-In Dangers of the American Automobile*. New York: Grossman.

Nagelhout, Ed, Julie Staggers, and Denise Tillery. 2009. "Risk Communication, Space, and Findability in the Public Sphere: A Case Study of a Physical and Online Information Center." *Journal of Technical Writing and Communication* 39.3: 227–243.

Nardi, B. A. and S. Whittaker. 2002. "The Place of Face-to-Face Communication in Distributed Work." In *Distributed Work: New Research on Working across Distance Using Technology*, edited by P. Hinds and Sara Kiesler, 83–110. Cambridge, MA: MIT University Press.

Nardi, B. A., S. Whittaker, and H. Schwarz. 2002. "NetWORKers and Their Activity in Intensional Networks." *Computer Supported Cooperative Work* 11: 205–242.

Neale, Dennis C. and John M. Carroll. 1997. "The Role of Metaphors in Interface Design." In *Handbook of Human-Computer Interaction*, edited by M. G. Helander, T. K. Landauer, and P. V. Prabhu, 441–462. Amsterdam: Elsevier Science Publishers.

Neil, Dan. 2007. "The 50 Worst Cars of All Time." *Time*. Last accessed June 16, 2014. http://content.time.com/time/specials/2007/article/0,28804,1658545_1658544_1658540,00.html.

———. 2008. "Ford Fusion Hybrid Intrudes on Despair." *Los Angeles Times*, December 19. http://articles.latimes.com/2008/dec/19/business/fi-neil19.

Nelson, T. H. 1990. "The Right Way to Think about Software Design." In *The Art of Human Computer Interface Design*, edited by B. Laurel, 235–243. Reading, MA: Addison Wesley.

Nissan. 2013. "Nissan LEAF with Highly Advanced Driver Assist System Gets First License Plate for Public Road Testing in Japan." *NissanNews.com*. Last modified September 26. http://nissannews.com/en-US/nissan/usa/releases/nissan-leaf-with-highly-advanced-driver-assist-system-gets-first-license-plate-for-public-road-testing-in-japan.

Oge, Margo T. 2015. *Driving the Future: Combating Climate Change with Cleaner, Smarter Cars*. New York: Arcade Publishing.

Packard, Vance. 1957. *The Hidden Persuaders*. New York: D. McKay & Co.

Packer, Jeremy and Stephen Crofts Wiley. 2011. *Communication Matters: Materialist Approaches to Media, Mobility and Networks*. New York: Routledge.

Pagé, Victor Wilfred. 1915. *The Model T Ford Car, Its Construction, Operation and Repair: A Complete Practical Treatise Explaining the Operating Principles of All Parts of the Ford Automobile, with Complete Instructions for Driving and Maintenance*. New York: Norman W. Henley.

Pender, Kelly. 2011. *Techne, from Neoclassicism to Postmodernism: Understanding Writing as a Useful, Teachable Art*. Anderson, SC: Parlor Press.

Peters, Björn and Lena Nilsson. 2007. "Modelling the Driver in Control." In *Modelling Driver Behaviour in Automotive Environments*, edited by P. Carlo Cacciabue, 85–104. London: Springer-Verlag.

Peters, Francis E. 1970. *Greek Philosophical Terms: A Historical Lexicon*. New York: New York University Press.

Pflugfelder, Ehren Helmut. 2009. "Something Less Than a Driver: Toward an Understanding of Gendered Bodies in Motorsport." *Journal of Sport and Social Issues* 33.4: 411–426.

———. 2011. "Sustaining Knowledge Work in Student Engineering Clubs." *Proceedings of the 2011 International Professional Communication Conference*. n.p.

———. 2015a. "Is No One at the Wheel?: Nonhuman Agency and Agentive Movement." In *Thinking with Bruno Latour in Rhetoric and Composition*, edited by Nathaniel Rivers and Paul Lynch, 176–200. Carbondale, IL: Southern Illinois University Press.

———. 2015b. "Rhetoric's New Materialism: From Micro-Rhetoric to Microbrew." *Rhetoric Society Quarterly* 45.5: 441–461.

———. 2015c. "Cell Phones, Networks, and Power: Documenting Cell Phone Literacies." *Kairos: A Journal of Rhetoric, Technology, and Pedagogy* 19.2. http://kairos.technorhetoric.net/19.2/topoi/pflugfelder/index.html.

Piaget, J. and B. Inhelder. 1967. *A Child's Conception of Space*. Translated by F. J. Langdon and J. L. Lunzer. New York: Norton.

Pigg, Stacey. 2014. "Coordinating Constant Invention: Social Media's Role in Distributed Work." *Technical Communication Quarterly* 23.2: 69–87.

Plato. 1997. In *Plato: Complete Works*, edited by John M. Cooper. Cambridge: Hackett Publishing.

Porter, James E. 2009. "Recovering Delivery for Digital Rhetoric." *Computers and Composition* 26.4: 207–224.

Potts, Liza. 2008 "Diagramming with Actor-Network Theory: A Method for Modeling Holistic Experience." *Proceedings of the 2008 International Professional Communication Conference*. n.p.

———. 2009. "Using Actor Network Theory to Trace and Improve Multimodal Communication Design." *Technical Communication Quarterly* 18.3: 281–301.

———. 2010. "Consuming Digital Rights: Mapping the Artifacts of Entertainment." *Technical Communication* 57.3: 300–318.

———. 2011. "Contextualizing Experiences: Tracing the Relationships between People and Technologies in the Social Web." *Journal of Business and Technical Communication* 77: 7–30.

———. 2013. *Social Media in Disaster Response*. New York: Routledge.

Potts, Liza and Angela Harrison. 2013. "Interfaces as Rhetorical Constructions: Reddit and 4chan During the Boston Marathon Bombings." In *SIGDOC '13. Proceedings of the 31*st *ACM International Conference on Design of Communication*, 143–150. New York: ACM.

Poulakos, John. 2002. "Kairos in Gorgias' Rhetorical Compositions." In *Rhetoric and Kairos: Essays in History, Theory, and Praxis*, edited by P. Sipora and J. S. Baumlin, 89–96. Albany, NY: State University of New York Press.

Pullman, George and Baotong Gu, eds. 2008. *Content Management: Bridging the Gap between Theory and Practice*. Amityville, NY: Baywood.

Ranney, T. 1994. "Models of Driving Behavior: A Review of Their Evolution." *Accident Analysis and Prevention* 26.6: 773–750.

Rasmussen, J. 1983. "Skills, Rules, and Knowledge: Signals, Signs, and Symbols, and Other Distinctions in Human Performance Models." *IEEE Transactions on Systems, Man, and Cybernetics, SMC* 13: 257–266.

Redish, Janice. 1995. "Adding Value as a Professional Technical Communicator." *Technical Communication* 42.1: 26–39.

———. 2010. "Technical Communication and Usability: Intertwined Strands and Mutual Influences." *IEEE Transactions on Professional Communication* 53.3: 191–201.

Redondo, Miguel, Crescencio Bravo, and Manuel Ortega, eds. 2009. *Engineering the User Interface: From Research to Practice*. London: Springer-Verlag.

Reich, Robert B. 1991. *The Work of Nations: Preparing Ourselves for 21st Century Capitalism*. New York: Knopf.

Reid, Alex. 2012. "Composing Objects: Prospects for a Digital Rhetoric #cwcon." *Digital Digs*, May 20. http://alex-reid.net/2012/05/composing-objects-prospects-for-a-digital-rhetoric-cwcon.html.

Revenge of the Electric Car. 2011. Directed by Chris Paine. Culver City, CA: Sony Pictures Home Entertainment. 2006. DVD.

Rice, Jeff. 2009. "Woodward Paths: Motorizing Space." *Technical Communication Quarterly* 18.3: 224–241.

———. 2012. *Digital Detroit: Rhetoric and Space in the Age of the Network*. Carbondale, IL: Southern Illinois University Press.

Rickert, Thomas. 2013. *Ambient Rhetoric: The Attunements of Rhetorical Being*. Pittsburgh, PA: University of Pittsburgh Press.

Ricoeur, Paul. 1981. *The Rule of Metaphor: Multi-Disciplinary Studies of the Creation of Meaning in Language*. Toronto: University of Toronto Press.

Ridolfo, Jim and Dànielle Nicole DeVoss. 2009. "Composing for Recomposition: Rhetorical Velocity and Delivery." *Kairos* 13.2: n.p. http://kairos.technorhetoric.net/13.2/topoi/ridolfo_devoss/intro.html.

Rieder, David M. 2013. "From GUI to NUI: Microsoft's Kinect and the Politics of the (Body as) Interface." *Present Tense: A Journal of Rhetoric in Society* 3.1: n.p. www.presenttensejournal.org/volume-3/from-gui-to-nui-microsofts-kinekt-and-the-politics-of-the-body-as-interface/.

Rittel, Horst W. J. and Melvin M. Webber. 1973. "Dilemmas in a General Theory of Planning." *Policy Sciences* 4: 155–169.

Roochnik, David. 1996. *Of Art and Wisdom: Plato's Understanding of Techne*. University Park, PA: Penn State University Press.

Rorty, Richard. 1989. *Contingency, Irony, and Solidarity*. Cambridge: Cambridge University Press.

Rubin, Eli. 2009. "The Trabant: Consumption, Eigen-Sinn, and Movement." *History Workshop Journal* 68.1: 27–44.

Rude, Carolyn D. 2004. "Toward an Expanded Concept of Rhetorical Delivery: The Uses of Reports in Public Policy Debates." *Technical Communication Quarterly* 13.3: 271–288.

Russell, David R. 1995. "Activity Theory and Its Implications for Writing Instruction." In *Reconceiving Writing, Rethinking Writing Instruction*, edited by Joseph Petraglia, 51–78. Hillsdale, NJ: Erlbaum.

———. 1997. "Rethinking Genre in School and Society: An Activity Theory Analysis." *Written Communication* 14.4: 504–554.

Salvo, Michael J. 2001. "Ethics of Engagement: User-Centered Design and Rhetorical Methodology." *Technical Communication Quarterly* 10.3: 273–290.

———. 2004. "Rhetorical Action in Professional Space: Information Architecture as Critical Practice." *Journal of Business and Technical Communication* 18.1: 39–66.

———. 2014. "What's in a Name?: Experience Architecture Rearticulates the Humanities." *Communication Design Quarterly* 3.2: 6–9.

Salvo, Michael J., Ehren Helmut Pflugfelder, and Joshua Prenosil. 2010. "The Children of Aramis." *Journal of Technical Writing and Communication* 40.3: 245–263.

Sauer, Beverly A. 2002. *The Rhetoric of Risk: Technical Documentation in Hazardous Environments*. Mahwah, NJ: Lawrence Erlbaum.

Savage, Gerald J. 2004. "Tricksters, Fools, and Sophists: Technical Communication as a Postmodern Rhetoric." In *Power and Legitimacy in Technical Communication, Vol. 2: Strategies for Professional Status*, edited by T. Kynell-Hunt and G. J. Savage, 167–193. Amityville, NY: Baywood.

Schoettle, Brandon and Michael Sivak. 2014. "Public Opinion About Self-Driving Vehicles in China, India, Japan, The U.S., The U.K., and Australia: UMTRI-2014–30." *University of Michigan Transportation Research Institute*. http://deepblue.lib.umich.edu/bitstream/handle/2027.42/108384/103024.pdf.

Schön, Donald. 1983. *The Reflective Practitioner: How Professionals Think in Action*. New York: Basic Books.

Schriver, Karen A. 1997. *Dynamics in Document Design: Creating Texts for Readers*. Hoboken, NJ: John Wiley.

Schultz, Jonathan. 2010. "G.M. Is Trying to Corner the Market on 'Range Anxiety.'" *New York Times*, September 2. http://wheels.blogs.nytimes.com/2010/09/02/g-m-is-trying-to-corner-the-market-on-range-anxiety/.

Schweinsberg. Christie. 2010. "Toyota Examines Prius Cruise Control Complaints." *WardsAuto*. Last updated March 22. http://wardsauto.com/ar/toyota_prius_cruise_100322.

Scientific American Frontiers. 1994. "Special from Germany." *Episode 402*, January 19. Narrated by Alan Alda. PBS. www.pbs.org/saf/transcripts/transcript402.htm#4.

Scott, J. Blake. 2003. *Risky Rhetoric: AIDS and the Cultural Practices of HIV Testing*. Carbondale, IL: Southern Illinois University Press.

Seifert, Jan. 2012. *IMG_3253*. Flickr, January 21. www.flickr.com/photos/58978138@N00/6770579011.

Selber, Stuart. 2010. "A Rhetoric of Electronic Instruction Sets." *Technical Communication Quarterly* 19.2: 95–117.

Selfe, Cynthia L. and Richard J. Selfe. 1994. "The Politics of the Interface: Power and Its Exercise in Electronic Contact Zones." *College Composition and Communication* 45.4: 480–504.

Sheller, Mimi. 2004. "Automotive Emotions: Feeling the Car." *Theory, Culture & Society* 21.4/5: 221–242.

Sheller, Mimi and John Urry. 2006. "The New Mobilities Paradigm." *Environment and Planning A* 38: 207–226.

Shoup, Donald. 2005. *The High Cost of Free Parking*. Chicago, IL: APA Planners Press.

Simon, Herbert A. 1969. *The Sciences of the Artificial*. Cambridge, MA: MIT University Press.

SINTEF. 2013. "Electric Transport with Wind in Its Sails." *ScienceDaily*, September 13. www.sciencedaily.com/releases/2013/09/130913085706.htm.

Sivak, Michael and Brandon Schoettle. 2015. *Motion Sickness in Self-Driving Vehicles, UMTRI-2015–12*. University of Michigan Transportation Research Institute. www.umich.edu/~umtriswt/PDF/UMTRI-2015–12_Abstract_English.pdf.

Slattery, Shaun. 2007. "Undistributing Work Through Writing: How Technical Writers Manage Texts in Complex Information Environments." *Technical Communication Quarterly* 16.3: 311–325.

Smiley.toerist. 2010. *Boulevard Victor Experimentation*. Wikimedia Commons, January 13. https://commons.wikimedia.org/wiki/File:Boulevard_Victor_experimentation.jpg.

Sneldon, Brian. 2010. *Trabant*. From Flickr, October 13. www.flickr.com/photos/exfordy/5088773103/.

Sperling, Daniel and Deborah Gordon. 2009. *Two Billion Cars: Driving Toward Sustainability*. Oxford: Oxford University Press.

Spinuzzi, Clay. 2001. " 'Light Green Doesn't Mean Hydrology!': Toward a Visual-Rhetorical Framework for Interface Design." *Computers and Composition* 18: 39–53.

———. 2003. *Tracing Genres through Organizations: A Sociocultural Approach to Information Design*. Cambridge, MA: MIT University Press.

———. 2004. "Describing Assemblages: Genre Sets, Systems, Repertoires, and Ecologies." Last updated May 5, 2004. www.dwrl.utexas.edu/content/describing-assemblages.

———. 2005. "Lost in the Translation: Shifting Claims in the Migration of a Research Technique." *Technical Communication Quarterly* 14.4: 411–446.

———. 2007. "Guest Editor's Introduction: Technical Communication in an Age of Distributed Work." *Technical Communication Quarterly* 16.3: 265–277.

———. 2008. *Network: Theorizing Knowledge Work in Telecommunications*. Cambridge: Cambridge University Press.

Srivastava, Mehul. 2010. "Apple Co-Founder Says His Prius Has Acceleration Flaw (Update1)." *Bloomberg*, February 3. www.bloomberg.com/apps/news?pid= newsarchive&sid=aVBaNu1.SHdw&pos=13.

Sullivan, Patricia. 1992. "Computer-Aided Publishing: Focusing on Documents." *Computers and Composition* 9.1: 135–149.

———. 1989/2009. "Human-Computer Interaction Perspectives on Word-Processing Issues." In *Rhetorically Rethinking Usability: Theories, Practices and Methodologies*, edited by Susan K. Miller-Cochran and Rochelle L. Roderigo, 15–33. Cresskill, NJ: Hampton Press.

Sun, Huatong. 2006. "The Triumph of Users: Achieving Cultural Usability Goals with User Localization." *Technical Communication Quarterly* 15.4: 457–481.

Swarts, Jason. 2007. "Mobility and Composition: The Architecture of Coherence in Non-Places." *Technical Communication Quarterly* 16.3: 279–309.

———. 2008. "Information Technologies as Discursive Agents: Methodological Implications for the Empirical Study of Knowledge Work." *Journal of Technical Writing and Communication* 38.4: 301–329.

———. 2010. "Recycled Writing: Assembling Actor Networks from Reusable Content." *Journal of Business and Technical Communication* 24.2: 127–163.

———. 2011. "Technological Literacy as Network Building." *Technical Communication Quarterly* 20.3: 274–302.

———. 2015. "Help Is in the Helping: An Evaluation of Help Documentation in a Networked Age." *Technical Communication Quarterly* 24.2: 164–187.

Teetor, Ralph R. 1950. *Speed Control Device for Resisting Operation of the Accelerator.* US Patent 2,519,859, filed August 11, 1948, issued August 22, 1950.

Tesla Motors. n.d. "Supercharger." Last accessed August 28, 2014. www.teslamotors.com/supercharger.

———. 2015. "Model S Software Update 6.2." March 19. http://my.teslamotors.com/de_AT/forum/forums/model-s-software-update-62-range-assurance-lte-valet-mode-blind-spot-and-more.

Thrift, Nigel. 1996. *Spatial Formations*. London: Sage Publications.

———. 2004. "Movement-Space: The Changing Domain of Thinking Resulting from the Development of New Kinds of Spatial Awareness." *Economy and Society* 33.4: 582–604.

———. 2008. *Non-Representational Theory: Space, Politics, Affect*. London: Routledge.

Thrun, Sebastian. 2002. "Probabilistic Robotics." *Communications of the ACM* 45.3: 52–57.

Thrun, Sebastian, Wolfram Burgard, and Deiter Fox. 2005. *Probabilistic Robotics*. Cambridge, MA: MIT University Press.

Thrun, Sebastian, Mike Montemerlo, Hendrik Dahlkamp, David Stavens, Andrei Aron, James Diebel, Philip Fong, John Gale, Morgan Halpenny, Gabriel Hoffmann, Kenny Lau, Celia Oakley, Mark Palatucci, Vaughan Pratt, Pascal Stang, Sven Strohband, Cedric Dupont, Lars-Erik Jendrossek, Christian Koelen, Charles Markey, Carlo Rummel, Joe van Niekerk, Eric Jensen, Philippe Alessandriniet, Gary Bradski, Bob Davies, Scott Ettinger, Adrian Kaehler, Ara Nefianal, and Pamela Mahoney. 2006. "Stanley: The Robot That Won the DARPA Grand Challenge." *Journal of Field Robotics* 23.9: 661–692.

Tidwell, Jenifer. 2005. *Designing Interfaces: Patterns for Effective Interaction Design*. Sebastapol, CA: O'Reilly Media.

Tolman, Edward C. 1948. "Cognitive Maps in Rats and Men." *The Psychological Review* 55.4: 189–208.

Trimble, Tyghe. 2010. "Ford's My Touch System Bring WiFi and Apps to Cars." *Popular Mechanics*, March 11. www.popularmechanics.com/technology/gadgets/news/4341692.

Trimbur, John. 2000. "Composition and the Circulation of Writing." *College Composition and Communication* 52.2: 188–219.

Trowbridge, C. C. 1913. "On Fundamental Methods of Orientation and 'Imaginary Maps.'" *Science* 19: 888–897.

Tversky, Barbara. 1993. "Cognitive Maps, Cognitive Collages, and Spatial Mental Models." In *Spatial Information Theory: A Theoretical Basis for GIS, Proceedings COSIT*

'93, Lecture Notes in Computer Science, edited by A. U. Frank and I. Campari, 14–24. Berlin: Springer.

United States Department of Commerce. 2011. "Commuting in the United States, 2009." Last accessed October, 2014, www.census.gov/prod/2011pubs/acs-15.pdf.

United States Department of State. 2010. *U.S. Climate Action Report.* www.state.gov/documents/organization/140636.pdf.

United States Department of Transportation. n.d. "Transportation's Role in Climate Change," *Transportation and Climate Change Clearinghouse.* Last accessed April 9, 2013, http://climate.dot.gov/about/transportations-role/overview.html.

United States Department of Transportation. Federal Highway Administration. 2013. "Cooperative Adaptive Cruise Control: Human Factors Analysis, FHWA-HRT-13–045." *FWHA.gov.* Last accessed October 22, 2014. www.fhwa.dot.gov/publications/research/safety/13045/13045.pdf.

United States Department of Transportation. National Highway Traffic Safety Administration. 2015. "Critical Reasons for Crashes Investigated in the National Motor Vehicle Crash Causation Survey, DOT HS 812 115." NRD.NHTSA.DOT.GOV. Last accessed March 22, 2015. www-nrd.nhtsa.dot.gov/pubs/812115.pdf.

United States Environmental Protection Agency. 2014. "Sources of Greenhouse Gas Emissions." *Epa.gov.* Last updated August 20, 2014, last accessed April 28, 2014. www.epa.gov/climatechange/ghgemissions/sources.html.

Urban, Tim. 2015. "How Tesla Will Change the World." *Wait but Why?* June. http://waitbutwhy.com/2015/06/how-tesla-will-change-your-life.html.

Urry, John. 2004. "The 'System' of Automobility." *Theory, Culture & Society* 21.4/5: 25–39.

———. 2006. "Inhabiting the Car." In *Against Automobility,* edited by Steffen Böhm, Campbell Jones, Chris Land, and Matthew Paterson, 17–31. Oxford: Wiley-Blackwell.

———. 2007. *Mobilities.* Cambridge: Polity.

Vanderbilt, Tom. 2008. *Traffic: Why We Drive the Way We Do.* New York: Knopf.

———. 2012. "The Single Most Important Object in the Global Economy." *Slate,* August 14. www.slate.com/articles/business/transport/2012/08/pallets_the_single_most_important_object_in_the_global_economy_.html.

Vatz, Richard E. 1973. "The Myth of the Rhetorical Situation." *Philosophy and Rhetoric* 6.3: 154–161.

Velte, Charles E. 1990. "Face Up to the Interface." *IEEE Transactions on Professional Communication* 33.3: 115–117.

Vinci Autoroutes. 2013. "Cruise Control and Speed Limiters Impact Driver Vigilance." July 30. http://fondation.vinci-autoroutes.com/fr/system/files/pdf/2013/07/pr_vinci_autoroutes_foundation_cruise_control_and_speed_limiters_impact_.pdf.

Virilio, Paul. 1995. *The Art of the Motor.* Minneapolis, MN: University of Minnesota Press.

Walters, Helen. 2010. "Ford and IDEO Redesign and Dashboard." *Businessweek.* Last modified January 7. www.businessweek.com/innovate/next/archives/2010/01/ford_and_ideo _redesign_the_dashboard.html.

Warnick, Barbara. 1993. *The Sixth Canon: Belletristic Rhetorical Theory and Its French Antecedents.* Columbia, SC: University of South Carolina Press.

Weir, D. H. and D. T. McRuer. 1968. "A Theory for Driver Steering Control of Motor Vehicles." *Road User Characteristics, Highway Research Record,* nr. 247: 7–28.

Weir, David H. and Kevin C. Chao. 2007. "Review of Control Theory Models for Directional and Speed Control." In *Modelling Driver Behaviour in Automotive Environments,* edited by P. Carlo Cacciabue, 293–312. London: Springer-Verlag.

Weisman, Jonathan. 2005. "Biggest Automaker Needs Big Changes." *Washington Post,* June 11. www.washingtonpost.com/wp-dyn/content/article/2005/06/10/AR200506100 2188.html.

Who Killed the Electric Car? 2006. Directed by Chris Paine. Culver City, CA: Sony Pictures Home Entertainment. 2006. DVD.

Wild, John. 1941. "Plato's Theory of Texnh: A Phenomenological Interpretation." *Philosophy and Phenomenological Research* 1: 255–293.

Winsor, Dorothy. 2006. "Using Writing to Structure Agency: An Examination of Engineers' Practice." *Technical Communication Quarterly* 15.4: 411–430.

Wysocki, Anne Frances and Julia I. Jasken. 2004. "What Should Be an Unforgettable Face . . ." *Computers and Composition* 21: 29–48.

Yates, J. and W. Orlikowski. 2002. "Genre Systems: Structuring Interaction through Communicative Norms." *Journal of Business Communication* 39.1: 13–35.

Zipcar, Inc. 2013. "Millenials and Technology." Last updated February 27, 2013, last accessed May 12, 2013. www.slideshare.net/Zipcar_Inc/millennial-slide-share-final-16812323.

Index

For Product Safety Concerns and Information please contact our EU representative GPSR@taylorandfrancis.com Taylor & Francis Verlag GmbH, Kaufingerstraße 24, 80331 München, Germany